Applied Natural Language Processing with Python

Implementing Machine Learning and Deep Learning Algorithms for Natural Language Processing

Taweh Beysolow II

Apress®

Applied Natural Language Processing with Python

Taweh Beysolow II
San Francisco, California, USA

ISBN-13 (pbk): 978-1-4842-3732-8   ISBN-13 (electronic): 978-1-4842-3733-5
https://doi.org/10.1007/978-1-4842-3733-5

Library of Congress Control Number: 2018956300

Copyright © 2018 by Taweh Beysolow II

Managing Director, Apress Media LLC: Welmoed Spahr
Acquisitions Editor: Celestin Suresh John
Development Editor: Siddhi Chavan
Coordinating Editor: Divya Modi

Cover designed by eStudioCalamar

Cover image designed by Freepik (www.freepik.com)

Distributed to the book trade worldwide by Springer Science+Business Media New York, 233 Spring Street, 6th Floor, New York, NY 10013. Phone 1-800-SPRINGER, fax (201) 348-4505, e-mail orders-ny@springer-sbm.com, or visit www.springeronline.com. Apress Media, LLC is a California LLC and the sole member (owner) is Springer Science + Business Media Finance Inc (SSBM Finance Inc). SSBM Finance Inc is a **Delaware** corporation.

For information on translations, please e-mail rights@apress.com, or visit http://www.apress.com/rights-permissions.

Apress titles may be purchased in bulk for academic, corporate, or promotional use. eBook versions and licenses are also available for most titles. For more information, reference our Print and eBook Bulk Sales web page at http://www.apress.com/bulk-sales.

Any source code or other supplementary material referenced by the author in this book is available to readers on GitHub via the book's product page, located at www.apress.com/978-1-4842-3732-8. For more detailed information, please visit http://www.apress.com/source-code.

Printed on acid-free paper

To my family, friends, and colleagues for their continued support and encouragement to do more with myself than I often can conceive of doing

Table of Contents

About the Author

Taweh Beysolow II is a data scientist and author currently based in San Francisco, California. He has a bachelor's degree in economics from St. Johns University and a master's degree in applied statistics from Fordham University. His professional experience has included working at Booz Allen Hamilton, as a consultant and in various startups as a data scientist, specifically focusing on machine learning. He has applied machine learning to federal consulting, financial services, and agricultural sectors.

About the Technical Reviewer

Santanu Pattanayak currently works at GE Digital as a staff data scientist and is the author of the deep learning book *Pro Deep Learning with TensorFlow: A Mathematical Approach to Advanced Artificial Intelligence in Python* (Apress, 2017). He has more than eight years of experience in the data analytics/data science field and a background in development and database technologies. Prior to joining GE, Santanu worked at companies such as RBS, Capgemini, and IBM. He graduated with a degree in electrical engineering from Jadavpur University, Kolkata, and is an avid math enthusiast. Santanu is currently pursuing a master's degree in data science from the Indian Institute of Technology (IIT), Hyderabad. He also devotes his time to data science hackathons and Kaggle competitions, where he ranks within the top 500 across the globe. Santanu was born and brought up in West Bengal, India, and currently resides in Bangalore, India, with his wife.

Acknowledgments

A special thanks to Santanu Pattanayak, Divya Modi, Celestin Suresh John, and everyone at Apress for the wonderful experience. It has been a pleasure to work with you all on this text. I couldn't have asked for a better team.

Introduction

Thank you for choosing *Applied Natural Language Processing with Python* for your journey into natural language processing (NLP). Readers should be aware that this text should not be considered a comprehensive study of machine learning, deep learning, or computer programming. As such, it is assumed that you are familiar with these techniques to some degree. Regardless, a brief review of the concepts necessary to understand the tasks that you will perform in the book is provided.

After the brief review, we begin by examining how to work with raw text data, slowly working our way through how to present data to machine learning and deep learning algorithms. After you are familiar with some basic preprocessing algorithms, we will make our way into some of the more advanced NLP tasks, such as training and working with trained word embeddings, spell-check, text generation, and question-and-answer generation.

All of the examples utilize the Python programming language and popular deep learning and machine learning frameworks, such as scikit-learn, Keras, and TensorFlow. Readers can feel free to access the source code utilized in this book on the corresponding GitHub page and/or try their own methods for solving the various problems tackled in this book with the datasets provided.

CHAPTER 1

What Is Natural Language Processing?

Deep learning and machine learning continues to proliferate throughout various industries, and has revolutionized the topic that I wish to discuss in this book: natural language processing (NLP). NLP is a subfield of computer science that is focused on allowing computers to understand language in a "natural" way, as humans do. Typically, this would refer to tasks such as understanding the sentiment of text, speech recognition, and generating responses to questions.

NLP has become a rapidly evolving field, and one whose applications have represented a large portion of artificial intelligence (AI) breakthroughs. Some examples of implementations using deep learning are chatbots that handle customer service requests, auto-spellcheck on cell phones, and AI assistants, such as Cortana and Siri, on smartphones. For those who have experience in machine learning and deep learning, natural language processing is one of the most exciting areas for individuals to apply their skills. To provide context for broader discussions, however, let's discuss the development of natural language processing as a field.

© Taweh Beysolow II 2018
T. Beysolow II, *Applied Natural Language Processing with Python*,
https://doi.org/10.1007/978-1-4842-3733-5_1

The History of Natural Language Processing

Natural language processing can be classified as a subset of the broader field of speech and language processing. Because of this, NLP shares similarities with parallel disciplines such as computational linguistics, which is concerned with modeling language using rule-based models. NLP's inception can be traced back to the development of computer science in the 1940s, moving forward along with advances in linguistics that led to the construction of formal language theory. Briefly, formal language theory models language on increasingly complex structures and rules to these structures. For example, the alphabet is the simplest structure, in that it is a collection of letters that can form strings called *words*. A formal language is one that has a regular, context-free, and formal grammar. In addition to the development of computer sciences as a whole, artificial intelligence's advancements also played a role in our continuing understanding of NLP.

In some sense, the single-layer perceptron (SLP) is considered to be the inception of machine learning/AI. Figure 1-1 shows a photo of this model.

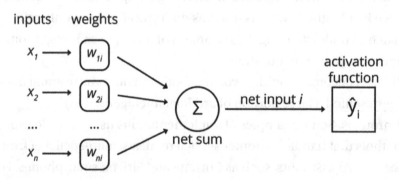

Figure 1-1. *Single-layer perceptron*

The SLP was designed by neurophysiologist Warren McCulloch and logician Walter Pitt. It is the foundation of more advanced neural network models that are heavily utilized today, such as multilayer perceptrons.

The SLP model is seen to be in part due to Alan Turing's research in the late 1930s on computation, which inspired other scientists and researchers to develop different concepts, such as formal language theory.

Moving forward to the second half of the twentieth century, NLP starts to bifurcate into two distinct groups of thought: (1) those who support a symbolic approach to language modelling, and (2) those who support a stochastic approach. The former group was populated largely by linguists who used simple algorithms to solve NLP problems, often utilizing pattern recognition. The latter group was primarily composed of statisticians and electrical engineers. Among the many approaches that were popular with the second group was Bayesian statistics. As the twentieth century progressed, NLP broadened as a field, including natural language understanding (NLU) to the problem space (allowing computers to react accurately to commands). For example, if someone spoke to a chatbot and asked it to "find food near me," the chatbot would use NLU to translate this sentence into tangible actions to yield a desirable outcome.

Skip closer to the present day, and we find that NLP has experienced a surge of interest alongside machine learning's explosion in usage over the past 20 years. Part of this is due to the fact that large repositories of labeled data sets have become more available, in addition to an increase in computing power. This increase in computing power is largely attributed to the development of GPUs; nonetheless, it has proven vital to AI's development as a field. Accordingly, demand for materials to instruct data scientists and engineers on how to utilize various AI algorithms has increased, in part the reason for this book.

Now that you are aware of the history of NLP as it relates to the present day, I will give a brief overview of what you should expect to learn. The focus, however, is primarily to discuss how deep learning has impacted NLP, and how to utilize deep learning and machine learning techniques to solve NLP problems.

A Review of Machine Learning and Deep Learning

You will be refreshed on important machine learning concepts, particularly deep learning models such as *multilayer perceptrons* (MLPs), *recurrent neural networks* (RNNs), and *long short-term memory* (LSTM) networks. You will be shown in-depth models utilizing toy examples before you tackle any specific NLP problems.

NLP, Machine Learning, and Deep Learning Packages with Python

Equally important as understanding NLP theory is the ability to apply it in a practical context. This book utilizes the Python programming language, as well as packages written in Python. Python has become the lingua franca for data scientists, and support of NLP, machine learning, and deep learning libraries is plentiful. I refer to many of these packages when solving the example problems and discussing general concepts.

It is assumed that all readers of this book have a general understanding of Python, such that you have the ability to write software in this language. If you are not familiar with this language, but you are familiar with others, the concepts in this book will be portable with respect to the methodology used to solve problems, given the same data sets. Be that as it may, this book is not intended to instruct users on Python. Now, let's discuss some of the technologies that are most important to understanding deep learning.

TensorFlow

One of the groundbreaking releases in open source software, in addition to machine learning at large, has undoubtedly been Google's TensorFlow. It is an open source library for deep learning that is a successor to Theano, a similar machine learning library. Both utilize data flow graphs for computational processes. Specifically, we can think of computations as dependent on specific individual operations. TensorFlow functionally operates by the user first defining a graph/model, which is then operated by a TensorFlow session that the user also creates.

The reasoning behind using a data flow graph rather than another computational format computation is multifaceted, however one of the more simple benefits is the ability to port models from one language to another. Figure 1-2 illustrates a data flow graph.

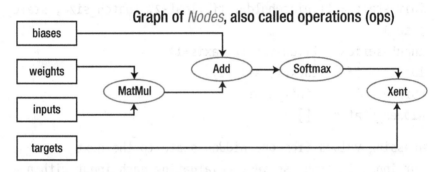

Figure 1-2. *Data flow graph diagram*

For example, you may be working on a project where Java is the language that is most optimal for production software due to latency reasons (high-frequency trading, for example); however, you would like to utilize a neural network to make predictions in your production system. Rather than dealing with the time-consuming task of setting up a training framework in Java for TensorFlow graphs, something could be written in Python relatively quickly, and then the graph/model could be restored by loading the weights in the production system by utilizing Java. TensorFlow code is similar to Theano code, as follows.

```
#Creating weights and biases dictionaries
weights = {'input': tf.Variable(tf.random_normal([state_
size+1, state_size])),
```

```
        'output': tf.Variable(tf.random_normal([state_size,
    n_classes]))}
biases = {'input': tf.Variable(tf.random_normal([1, state_
size])),
        'output': tf.Variable(tf.random_normal([1, n_classes]))}

#Defining placeholders and variables
X = tf.placeholder(tf.float32, [batch_size, bprop_len])
Y = tf.placeholder(tf.int32, [batch_size, bprop_len])
init_state = tf.placeholder(tf.float32, [batch_size, state_
size])
input_series = tf.unstack(X, axis=1)
labels = tf.unstack(Y, axis=1)
current_state = init_state
hidden_states = []

#Passing values from one hidden state to the next
for input in input_series: #Evaluating each input within
the series of inputs
    input = tf.reshape(input, [batch_size, 1]) #Reshaping
    input into MxN tensor
    input_state = tf.concat([input, current_state], axis=1)
    #Concatenating input and current state tensors
    _hidden_state = tf.tanh(tf.add(tf.matmul(input_
    state, weights['input']), biases['input'])) #Tanh
    transformation
    hidden_states.append(_hidden_state) #Appending the next
    state
    current_state = _hidden_state #Updating the current state
```

TensorFlow is not always the easiest library to use, however, as there often serious gaps between documentation for toy examples vs. real-world examples that reasonably walk the reader through the complexity of implementing a deep learning model.

In some ways, TensorFlow can be thought of as a language inside of Python, in that there are syntactical nuances that readers must become aware of before they can write applications seamlessly (if ever). These concerns, in some sense, were answered by Keras.

Keras

Due to the slow development process of applications in TensorFlow, Theano, and similar deep learning frameworks, Keras was developed for prototyping applications, but it is also utilized in production engineering for various problems. It is a wrapper for TensorFlow, Theano, MXNet, and DeepLearning4j. Unlike these frameworks, defining a computational graph is relatively easy, as shown in the following Keras demo code.

```python
def create_model():
    model = Sequential()
    model.add(ConvLSTM2D(filters=40, kernel_size=(3, 3),
                         input_shape=(None, 40, 40, 1),
                         padding='same', return_sequences=True))
    model.add(BatchNormalization())

    model.add(ConvLSTM2D(filters=40, kernel_size=(3, 3),
                         padding='same', return_sequences=True))
    model.add(BatchNormalization())

    model.add(ConvLSTM2D(filters=40, kernel_size=(3, 3),
                         padding='same', return_sequences=True))
    model.add(BatchNormalization())

    model.add(ConvLSTM2D(filters=40, kernel_size=(3, 3),
                         padding='same', return_sequences=True))
    model.add(BatchNormalization())

    model.add(Conv3D(filters=1, kernel_size=(3, 3, 3),
```

```
                  activation='sigmoid',
                  padding='same', data_format='channels_last'))
    model.compile(loss='binary_crossentropy', optimizer='adadelta')
    return model
```

Although having the added benefit of ease of use and speed with respect to implementing solutions, Keras has relative drawbacks when compared to TensorFlow. The broadest explanation is that Keras users have considerably less control over their computational graph than TensorFlow users. You work within the confines of a sandbox when using Keras. TensorFlow is better at natively supporting more complex operations, and providing access to the most cutting-edge implementations of various algorithms.

Theano

Although it is not covered in this book, it is important in the progression of deep learning to discuss Theano. The library is similar to TensorFlow in that it provides developers with various computational functions (add, matrix multiplication, subtract, etc.) that are embedded in tensors when building deep learning and machine learning models. For example, the following is a sample Theano code.

```
(code redacted please see github)
X, Y = T.fmatrix(), T.vector(dtype=theano.config.floatX)
    weights = init_weights(weight_shape)
    biases = init_biases(bias_shape)
    predicted_y = T.argmax(model(X, weights, biases), axis=1)

    cost = T.mean(T.nnet.categorical_crossentropy(predicted_y, Y))
    gradient = T.grad(cost=cost, wrt=weights)
    update = [[weights, weights - gradient * 0.05]]
```

```
    train = theano.function(inputs=[X, Y], outputs=cost,
updates=update, allow_input_downcast=True)
    predict = theano.function(inputs=[X], outputs=predicted_y,
allow_input_downcast=True)

    for i in range(0, 10):
        print(predict(test_x_data[i:i+1]))
if __name__ == '__main__':

    model_predict()
```

When looking at the functions defined in this sample, notice that T is the variable defined for a tensor, an important concept that you should be familiar with. Tensors can be thought of as objects that are similar to vectors; however, they are distinct in that they are often represented by arrays of numbers, or functions, which are governed by specific transformation rules unique unto themselves. Tensors can specifically be a single point or a collection of points in space-time (any function/model that combines x, y, and z axes plus a dimension of time), or they may be a defined over a continuum, which is a *tensor field*. Theano and TensorFlow use tensors to perform most of the mathematical operations as data is passed through a computational graph, colloquially known as a *model*.

It is generally suggested that if you do not know Theano, you should focus on mastering TensorFlow and Keras. Those that are familiar with the Theano framework, however, may feel free to rewrite the existing TensorFlow code in Theano.

Applications of Deep Learning to NLP

This section discusses the applications of deep learning to NLP.

Introduction to NLP Techniques and Document Classification

In Chapter 3, we walk through some introductory techniques, such as word tokenization, cleaning text data, term frequency, inverse document frequency, and more. We will apply these techniques during the course of our data preprocessing as we prepare our data sets for some of the algorithms reviewed in Chapter 2. Specifically, we focus on classification tasks and review the relative benefits of different feature extraction techniques when applied to document classification tasks.

Topic Modeling

In Chapter 4, we discuss more advanced uses of deep learning, machine learning, and NLP. We start with topic modeling and how to perform it via latent Dirichlet allocation, as well as non-negative matrix factorization. Topic modeling is simply the process of extracting topics from documents. You can use these topics for exploratory purposes via data visualization or as a preprocessing step when labeling data.

Word Embeddings

Word embeddings are a collection of models/techniques for mapping words (or phrases) to vector space, such that they appear in a high-dimensional field. From this, you can determine the degree of similarity, or dissimilarity, between one word (or phrase, or document) and another. When we project the word vectors into a high-dimensional space, we can envision that it appears as something like what's shown in Figure 1-3.

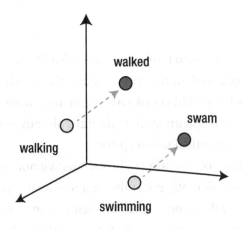

Figure 1-3. *Visualization of word embeddings*

Ultimately, how you utilize word embeddings is up to your own interpretation. They can be modified for applications such as spell check, but can also be used for sentiment analysis, specifically when assessing larger entities, such as sentences or documents in respect to one another. We focus simply on how to train the algorithms and how to prepare data to train the embeddings themselves.

Language Modeling Tasks Involving RNNs

In Chapter 5, we end the book by tackling some of the more advanced NLP applications, which is after you have been familiarized with preprocessing text data from various format and training different algorithms. Specifically, we focus on training RNNs to perform tasks such as name entity recognition, answering questions, language generation, and translating phrases from one language to another.

Summary

The purpose of this book is to familiarize you with the field of natural language processing and then progress to examples in which you can apply this knowledge. This book covers machine learning where necessary, although it is assumed that you have already used machine learning models in a practical setting prior.

While this book is not intended to be exhaustive nor overly academic, it is my intention to sufficiently cover the material such that readers are able to process more advanced texts more easily than prior to reading it. For those who are more interested in the tangible applications of NLP as the field stands today, it is the vast majority of what is discussed and shown in examples. Without further ado, let's begin our review of machine learning, specifically as it relates to the models used in this book.

CHAPTER 2

Review of Deep Learning

You should be aware that we use deep learning and machine learning methods throughout this chapter. Although the chapter does not provide a comprehensive review of ML/DL, it is critical to discuss a few neural network models because we will be applying them later. This chapter also briefly familiarizes you with TensorFlow, which is one of the frameworks utilized during the course of the book. All examples in this chapter use toy numerical data sets, as it would be difficult to both review neural networks and learn to work with text data at the same time.

Again, the purpose of these toy problems is to focus on learning how to create a TensorFlow model, *not* to create a deployable solution. Moving forward from this chapter, all examples focus on these models with text data.

Multilayer Perceptrons and Recurrent Neural Networks

Traditional neural network models, often referred to as *multilayer perceptron models* (MLPs), succeed *single-layer perceptron models* (SLPs). MLPs were created to overcome the largest shortcoming of the SLP model, which was the inability to effectively handle data that is not linearly separable. In practical cases, we often observe that multivariate data is

© Taweh Beysolow II 2018
T. Beysolow II, *Applied Natural Language Processing with Python*,
https://doi.org/10.1007/978-1-4842-3733-5_2

non-linear, rendering the SLP null and void. MLPs are able to overcome this shortcoming—specifically because MLPs have multiple layers. We'll go over this detail and more in depth while walking through some code to make the example more intuitive. However, let's begin by looking at the MLP visualization shown in Figure 2-1.

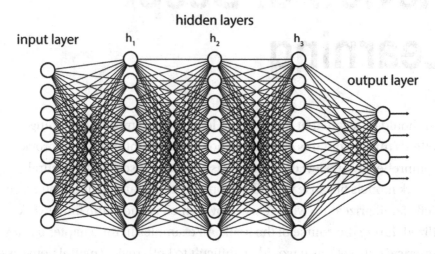

Figure 2-1. *Multilayer perceptron*

Each layer of an MLP model is connected by weights, all of which are initialized randomly from a standard normal distribution. The input layer has a set number of nodes, each representative of a feature within a neural network. The number of hidden layers can vary, but each of them typically has the same number of nodes, which the user specifies. In regression, the output layer has one node. In classification, it has K nodes, where K is the number of classes.

Next, let's have an in-depth discussion on how an MLP works and complete an example in TensorFlow.

Toy Example 1: Modeling Stock Returns with the MLP Model

Let's imagine that we are trying to predict Ford Motor Company (F) stock returns given the returns of other stocks on the same day. This is a regression problem, given that we are trying to predict a continuous value. Let's begin by defining an mlp_model function with arguments that will be used later, as follows:

```
def mlp_model(train_data=train_data, learning_rate=0.01,
iters=100, num_hidden1=256):
```

This Python function contains all the TensorFlow code that forms the body of the neural network. In addition to defining the graph, this function invokes the TensorFlow session that trains the network and makes predictions. We'll begin by walking through the function, line by line, while tying the code back to the theory behind the model.

First, let's address the arguments in our function: train_data is the variable that contains our training data; in this example; it is the returns of specific stocks over a given period of time. The following is the header of our data set:

```
0   0.002647 -0.001609  0.012800  0.000323  0.016132 -0.004664
-0.018598
1   0.000704  0.000664  0.023697 -0.006137 -0.004840  0.003555
-0.006664
2   0.004221  0.003600  0.002469 -0.010400 -0.008755 -0.002737
0.025367
3   0.003328  0.001605  0.050493  0.006897  0.010206  0.002260
-0.007156
4   0.001397  0.004052 -0.009965 -0.012720 -0.019235 -0.002255
0.017916
```

5 -0.009326 -0.003754 -0.014506 -0.006607 -0.034865 0.011463
0.003844
6 0.008446 0.005747 0.022830 0.009312 0.021757 -0.000319
0.023982
7 0.002705 0.002623 0.007636 0.020099 -0.007433 -0.008303
-0.004330
8 -0.011224 -0.009530 -0.008161 -0.003230 -0.015381 -0.003381
-0.010674
9 0.012496 0.010942 0.016750 0.007777 0.001233 0.008724
0.033367

Each of the columns represent the percentage return of a stock on a given day, with our training set containing 1180 observations and our test set containing 582 observations.

Moving forward, we come to the learning rate and activation function. In machine learning literature, the learning rate is often represented by the symbol η (eta). The learning rate is a scalar value that controls the degree to which the gradient is updated to the parameter that we wish to change. We can exemplify this technique when referring to the gradient descent update method. Let's first look at the equation, and then we can break it down iteratively.

$$\theta_{t+1} = \theta_t - \eta \frac{1}{N} \Sigma_{i=1}^{N} \left(h_\theta(x)^i - y^i \right)^2 \tag{2.1}$$

$$\theta_{t+1} = \theta_t - \eta \frac{1}{N} \Sigma_{i=1} 2\left(h_\theta(x)^i - y^i \right) \nabla_\theta h_\theta(x)^i$$

In Equation 2.1, we are updating some parameter, θ, at a given time step, t. $h_\theta(x)^i$ is equal to the hypothesized label/value, y being the actual label/value, in addition to N being equal to the total number of observations in the data set we are training on.

$\nabla_\theta h_\theta(x)^i$ is the gradient of the output with respect to the parameters of the model.

Each unit in a neural network (with the exception of the input layer) receives the weighted sum of inputs multiplied by weights, all of which are summed with a bias. Mathematically, this can be described in Equation 2.2.

$$y = f\left(x, w^{T}\right) + b \qquad (2.2)$$

In neural networks, the parameters are the weights and biases. When referring to Figure 2-1, the weights are the lines that connect the units in a layer to one another and are typically initialized by randomly sampling from a normal distribution. The following is the code where this occurs:

```
weights = {'input': tf.Variable(tf.random_normal([train_x.
shape[1], num_hidden])),
          'hidden1': tf.Variable(tf.random_normal([num_
          hidden, num_hidden])),
          'output': tf.Variable(tf.random_normal([num_hidden,
          1]))}

biases = {'input': tf.Variable(tf.random_normal([num_
hidden])),
          'hidden1': tf.Variable(tf.random_normal([num_
          hidden])),
          'output': tf.Variable(tf.random_normal([1]))}
```

Because they are part of the computational graph, weights and biases in TensorFlow must be initialized as TensorFlow variables with the tf. Variable(). TensorFlow thankfully has a function that generates numbers randomly from a normal distribution called tf.random_normal(), which takes an array as an argument that specifies the shape of the matrix that you are creating. For people who are new to creating neural networks, choosing the proper dimensions for the weight and bias units is a typical source of frustration. The following are some quick pointers to keep in mind :

- When referring to the weights, the columns of a given layer need to match the rows of the next layer.

- The columns of the weights for every layer must match the number of units for each layer in the biases.

- The output layer columns for the weights dictionary (and array shape for the bias dictionary) should be representative of the problem that you are modeling. (If regression, 1; if classification, N, where N = the number of classes).

You might be curious as to why we initialize the weights and biases randomly. This leads us to one of the key components of neural networks' success. The easiest explanation is to imagine the following two scenarios:

- **All weights are initialized to 1**. If all the weights are initialized to 1, every neuron is passed the same value, equal to the weighted sum, plus the bias, and then put into some activation function, whatever this value may be.

- **All weights are initialized to 0**. Similar to the prior scenario, all the neurons are passed the same value, except that this time, the value is definitely zero.

The more general problem associated with weights that are initialized at the same location is that it makes the network susceptible to getting stuck in local minima. Let's imagine an error function, such as the one shown in Figure 2-2.

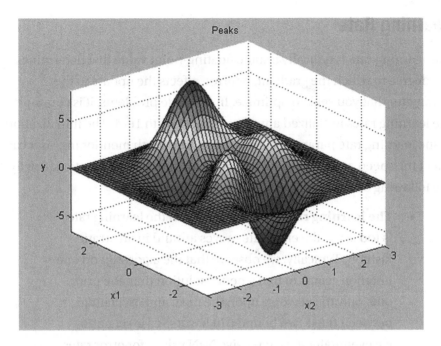

Figure 2-2. *Error plot*

Imagine that when we initialize the neural network weights at 0, and subsequently that when it calculates the error, it yields the value at the Y variable in Figure 2-2. The gradient descent algorithm always gives the same update for the weights from the first iteration of the algorithm, and it likely gives the same value moving forward. Because of that, we are not taking advantage of the ability of neural networks to start from any point in the solution space. This effectively removes the stochastic nature of neural networks, and considerably reduces the probability of reaching the best possible solution for the weight optimization problem. Let's discuss the learning rate.

Learning Rate

The learning rate is typically a static floating-point value that determines the degree to which the gradient, or error, affects the update to the parameter that you seek to optimize. In example problems, it is common to see learning rates initialized anywhere from 0.01 to 1e–4. The initialization of the learning rate parameter is another point worth mentioning, as it can affect the speed at which the algorithm converges upon a solution. Briefly, the following are two scenarios that are problematic:

- **The learning rate is too large.** When the learning rate is too large, the error rate moves around in an erratic fashion. Typically, we observe that the algorithm on one iteration seems to find a better solution than the prior one, only to get worse upon the next, and oscillating between these two bounds. In a worst-case scenario, we eventually start to receive NaN values for error rates, and all solutions effectively become defunct. This is the exploding gradient problem, which I discuss later.

- **The learning rate is too small.** Although, over time, this does not yield an incorrect, ultimately, we spend an inordinate amount of time waiting for the solution to converge upon an optimal solution.

The optimal solution is to pick a learning rate that is large enough to minimize the number of iterations needed to converge upon an optimal solution, while not being so large that it passes this solution in its optimization path. Some solutions, such as adaptive learning rate algorithms, solve the problem of having to grid search or iteratively use different learning rates. The mlp_model() function uses the Adam (**ada**ptive **m**oment estimation) optimization algorithm, which updates the learning rate aw we learn. I briefly discuss how this algorithm works, and why you should use it to expedite learning rate optimization.

Adam was first described in a paper that written by Diederik Kingma and Jimmy Lei Ba. Adam specifically seeks to optimize learning rates by estimating the first and second moments of the gradients. For those who are unfamiliar, *moments* are defined as specific measures of the shape of a set of points. As it relates to statistics, these points are typically the values within a probability distribution. We can define the zeroth moment as the total probability; the first moment as the mean; and second moment as the variance. In this paper, they describe the optimal parameters for Adam, in addition to some initial assumptions, as follows:

- α = Stepsize; $\alpha := 0.001$, $\epsilon = 10^{-8}$

- β_1, β_2 = Exponential decay rates for 1st and 2nd moment estimateions $\beta_1 = 0.9$, $\beta_2 = 0.999$; $\beta_1, \beta_2 \in [0, 1)$

- $f(\theta)$ = Stochastic objective function that we are optimizing with parameters θ

- m = 1st moment vector, v = 2nd moment vector (Both initialized as 0s)

With this in mind, although we have not converged upon an optimal solution, the following is the algorithm that we use:

- $g_t = \nabla_\theta f_t(\theta_{t-1})$

- $\hat{m}, \hat{v} = \text{Bias} - \text{corrected first and second moment estimates}$
  respectively;

- $m_t := \beta_1 * m_{t-1} + (1 - \beta_1) * g_t, v_t := \beta_2 * v_{t-1} + (1 - \beta_2) * g_t^2$

- $\hat{m}_t := \dfrac{m_t}{1 - \beta_1^t}, \hat{v}_t := \dfrac{v_t}{1 - \beta_2^t}$

- $\theta_t := \theta_{t-1} - \alpha * \dfrac{\hat{m}_t}{\sqrt{(\hat{v}_t)} + \epsilon}$

While the preceding formulae describe Adam when optimizing one parameter, we can extrapolate the formulae to adjust for multiple parameters (as is the case with multivariate problems). In the paper, Adam outperformed other standard optimization techniques and was seen as the default learning rate optimization algorithm.

As for the final parameters, num_hidden refers to the number of units in the hidden layer(s). A commonly referenced rule of thumb is to make this number equal to the number of inputs plus the number of outputs, and then multiplied by 2/3.

Epochs refers to the number of times the algorithm should iterate through the entirety of the training set. Given that this is situation dependent, there is no general suggestible number of epochs that a neural network should be trained. However, a suggestible method is to pick an arbitrarily large number (1500, for example), plot the training error, and then observe which number of epochs is sufficient. If needed, you can enlarge the upper limit to allow the model to optimize its solution further.

Now that I have finished discussing the parameters, let's walk through the architecture, code, and mathematics of the multilayer perceptron, as follows:

```
#Creating training and test sets
train_x, train_y = train_data[0:int(len(train_data)*.67),
1:train_data.shape[1]], train_data[0:int(len(train_data)*.67), 0]
test_x, test_y = train_data[int(len(train_data)*.67):, 1:train_
data.shape[1]], train_data[int(len(train_data)*.67):, 0]
```

Observe that we are creating both a training set and a test set. The training and test sets contain 67% and 33%, respectively, of the original data set labeled train_data. It is suggested that machine learning problems have these two data sets, at a minimum. It is optional to create a validation set as well, but this step is omitted for the sake of brevity in this example.

Next, let's discuss the following important aspect of working with TensorFlow:

```
#Creating placeholder values and instantiating weights and
biases as dictionaries
```

```
X = tf.placeholder('float', shape = (None, 7))
Y = tf.placeholder('float', shape = (None, 1))
```

When working in TensorFlow, it is important to refer to machine learning models as *graphs*, since we are creating computational graphs with different tensor objects. Any typical deep learning or machine learning model expects an explanatory and response variable; however, we need to specify what these are. Since they are not a part of the graph, but are representational objects that we are passing data through, they are defined as *placeholder variables*, which we can access from TensorFlow (imported as tf) by using tf.placeholder(). The three arguments for this function are dtype (data type), shape, and name. dtype and shape are the only required arguments. The following are quick rules of thumb:

- Generally, the shape of the X and Y variables should be initialized as a tuple. When working with a two-dimensional data set, the shape of the X variable should be (none, number of features), and the shape of the Y variable should be (none, [1 if regression, N if classification]), where N is the number of classes.

- The data type specified for these placeholders should reflect the values that you are passing through them. In this instance, we are passing through a matrix of floating-point values and predicting a floating-point value, so both placeholders for the response and explanatory variables have the float data type. In the instance that this was a classification problem, assuming the same data passed through the explanatory variable, the response variable has the int data type since the labels for the classes are integers.

Since I discussed the weights in the neural network already, let's get to the heart of the neural network structure: the input through the output layers, as shown in the following code (inside mlp_model() function):

```
#Passing data through input, hidden, and output layers
input_layer = tf.add(tf.matmul(X, weights['input']),
biases['input']) (1)
input_layer = tf.nn.sigmoid(input_layer) (2)
input_layer = tf.nn.dropout(input_layer, 0.20) (3)

hidden_layer = tf.add(tf.multiply(input_layer,
weights['hidden1']), biases['hidden1'])
hidden_layer = tf.nn.relu(hidden_layer)
hidden_layer = tf.nn.dropout(hidden_layer, 0.20)

output_layer = tf.add(tf.multiply(hidden_layer, weights
['output']),biases['output']) (4)
```

When looking at the first line of highlighted code (1), we see the input layer operation. Mathematically, operations from one neural network layer to the next can be represented by the following equation:

$$layer_k = f\left(\left(X_k * w_k^T\right) + bias_k\right) \tag{2.2.1}$$

$f(x)$ is equal to some activation function. The output from this operation is passed to the next layer, where the same operation is run, including any operations placed between layers. In TensorFlow, there are built-in mathematical operations to represent the preceding equation: tf.add() and tf.matmul().

After we create the output, which in this instance is a matrix of shape (1, 256), we pass it to an activation function. In the second line of highlighted code (2), we first pass the weighted sum of the inputs and bias to a sigmoid activation function, given in Equation 2.3.

$$\sigma = \left(\frac{1}{1 + e^{-x}}\right) \tag{2.3}$$

e is the exponential function. Activation functions serve as a way to scale the outputs from Equation 2.2, and are sometimes directly related to how we classify outputs. More importantly, this is the core component of the neural network that introduces non-linearity to the learning process. Simply stated, if we use a linear activation function, where $f(x) = x$, we are simply repetitively passing the outputs of a linear function from the input layer to the output layer. Figure 2-3 illustrates this activation function.

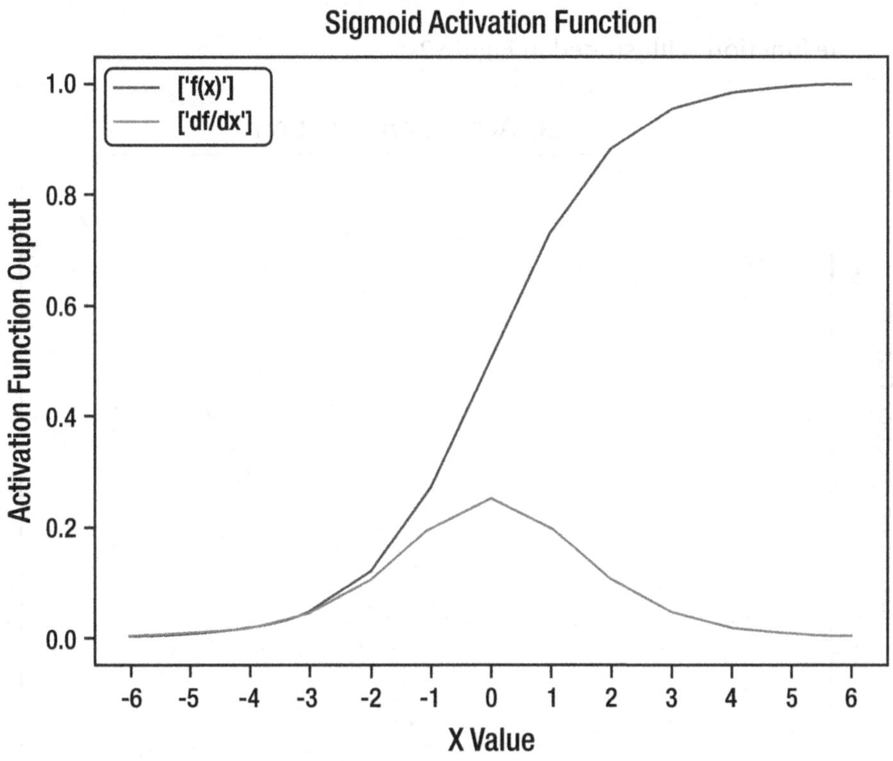

Sigmoid Activation Function

Figure 2-3. *Sigmoid activation function*

Although the range here is from –6 to 6, the function essentially looks like $-\infty$ to ∞, in that there are asymptotes at 0 and 1 as X grows infinitely larger or infinitely smaller, respectively. This function is one of the more common activation functions utilized in neural networks, which we use in the first layer.

Also, we defined the derivative of this function, which is important in mathematically explaining the vanishing gradient problem (discussed later in the chapter). Although going through all the activation functions in neural networks would be exhaustive, it is worth discussing the other activation function that this neural network utilizes. The hidden layer uses a ReLU activation function, which is mathematically defined in Equation 2.4.

$$ReLU(x) = \max(0, x) \tag{2.4}$$

The function is illustrated in Figure 2-4.

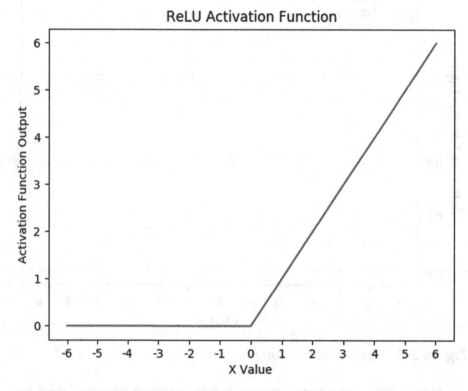

Figure 2-4. *ReLU activation function*

Both mathematically and visually, the ReLU activation function is simple. The output of a ReLU is a matrix of 0s, with some positive values. One of the major benefits of the ReLU activation function lies in the fact that it produces a sparse matrix as its output. This attribute is ultimately why I have decided to include it as the activation function in the hidden layer, particularly as it relates to the vanishing gradient problem.

Vanishing Gradients and Why ReLU Helps to Prevent Them

The vanishing gradient problem is specific to the training of neural networks, and it is part of the improvements that researchers sought to make with LSTM over RNN (both are discussed later in this chapter). The vanishing gradient problem is a phenomenon observed when the gradient gets so small that the updates to weights from one iteration to the next either stops completely or is considerably negligible.

Logically, what proceeds is a situation in which the neural network effectively stops training. In most cases, this results in poor weight optimization, and ultimately, bad training and test set performance. Why this happens can be explained precisely by how the updates for each of the weights are calculated:

When we look at Figure 2-3, we see the derivative of the sigmoid function. The majority of the function's derivate falls in a narrow range, with most of the values being close to 0. When considering how to calculate the gradient of differing hidden layers, this is precisely what causes a problem as our network gets deeper. Mathematically, this is represented by the following equation:

$$\frac{\partial E_3}{\partial W} = \sum_{k=0}^{2} \frac{\partial E_3}{\partial \widehat{y_3}} \frac{\partial y_3}{\partial s_3} \frac{\partial s_3}{\partial s_k} \frac{\partial s_k}{\partial W}$$

As you can see, when we backpropagate the error to layer k, which in this example is 0 (the input layer), we are multiplying several derivatives of the activation function's output several times. This is a brief explanation of the chain rule and underlies most of a neural networks' backpropagation training algorithm. The chain rule is a formula that specifies how to calculate a derivative that is composed of two or more functions. Assume that we have a two-layer neural network. Let's also assume that our respective gradients are 0.001 and 0.002. This yields 2 e–6 as a respective gradient of the output layer. Our update to the next gradient would be described as negligible.

You should know that any activation function that yields non-sparse outputs, particularly when used for multiple layers in succession, typically causes vanishing gradients. We are able to substantially mitigate this problem by using a combination of sparse and non-sparse output activation functions, or exclusively utilize non-spare activation functions. We illustrate an example of such a neural network in the `mlp_model()` function. For now, however, let's take a look at one last activation layer before we finish analyzing this MLP.

Observe that after every activation layer, we use the *dropout layer*, invoked by `tf.nn.dropout()`. Dropout layers have the same dimensions as the layer preceding them; however, they arbitrarily set a random selection of weights' values to 0, effectively "shutting off" the neurons that are connected to them. In every iteration, there are a different set of random neurons that shut off. The benefit of using dropout is to prevent overfitting, which is the instance in which a model performs well in training data but poorly in test data.

There are a multitude of factors that can cause overfitting, including (but not limited to) not having enough training data or not cross-validating data (which induces a model to memorize idiosyncrasies of a given orientation of a data set rather than generalizing to the distribution underlying the data). Although you should solve issues like these first, adding dropout is not a bad idea. When you execute functions without dropout, you notice overfitting relative to the models that do contain dropout.

Let's discuss some final MLP topics—specifically, the key components to what causes the model to learn.

Loss Functions and Backpropagation

Loss functions are specifically how we define the degree to which our model was incorrect. In regression, the most typical choices are *mean squared error* (MSE) or *root mean squared error* (RMSE). Mathematically, they are defined as follows:

$$MSE = \frac{1}{N} \sum_{i=1}^{N} \left(h_\theta(x_i) - y^i \right)^2 \tag{2.5}$$

$$RMSE = \sqrt{\frac{1}{N} \sum_{i=1}^{N} \left(h_\theta(x_i) - y^i \right)^2} \tag{2.6}$$

error = tf.reduce_mean(tf.pow(output_layer – Y,2)) (mean squared error in code)

Intuitively, MSE (see Equation 2.5) provides a method for assessing what was the average error over all predictions in a given epoch. RMSE (see Equation 2.6) provides the same statistic, but takes the square root of the MSE value. The benefit of RMSE is that it provides a statistic in the same unit as the predicted value, allowing the user to assess the performance of the model more precisely. MSE does not have this benefit, and as such, it becomes less interpretable—except in the sense that a lower MSE from one epoch to the next is desirable.

As an example, if we are predicting money, what does it mean that our prediction is $0.30 squared inaccurate? While we can tell that we have a better solution if the next epoch yields an MSE of $0.10, it is much harder to tell precisely what an MSE of $0.10 translates to in a given prediction. We compare the results of using RMSE vs. MSE in the final toy example in the chapter. In natural language processing, however, we more often deal with error functions reserved for classification tasks. With that in mind, you should be accustomed to the following formulas.

The binary cross entropy is

$$\mathcal{L}\big(y, h_x(\theta)\big)_i = -y\log(p) + (1-y)\log(1-p)) \tag{2.7}$$

The multiclass cross entropy is

$$\mathcal{L}\big(y, h_x(\theta)\big)_i = \max\big(0, s_j - s_{y_i} + \Delta\big) \tag{2.8}$$

Cross entropy is the number of bits needed to identify an event drawn from a set. The same principles (with respect to training using an MSE or RMSE loss function) are carried when using a cross-entropy-based loss function. Our objective is to optimize the weights in a direction that minimizes the error as much as possible.

At this point, we have walked through the MLP from the initialization of the parameters, what they mean, how the layer moves from each layer, what the activation functions do to it, and how the error is calculated. Next, let's dig into recurrent neural networks, long short-term memory, and their relative importance in the field of natural language processing.

Recurrent Neural Networks and Long Short-Term Memory

Despite the relative robustness of MLPs, they have their limitations. The model assumes independence between inputs and outputs, making it a suboptimal choice for problems in which the output of function is statistically dependent on the preceding inputs. As this relates to natural language processing (NLP), there are tasks that MLPs might be particularly useful for, such as sentiment analysis. In these problems, one body of text being classified as negative is not dependent on assessing the sentiment of a separate body of text.

As an example, I wouldn't need to read multiple restaurant reviews to determine whether an individual review is positive or negative. It can be determined by the attributes of a given observation. However, this is not always the type of NLP problem we encounter. For example, let's assume that we are trying to spell-check on the following sentences:

"I am happy that we are going too the mall!"

"I am happy to. That class was excellent."

Both sentences are incorrect in their usage of the words *too* and *to*, respectively, because of the context in which they appear. We must use the sequence of words prior, and perhaps even the words after, to determine what is incorrect. Another similar problem would be predicting words in a sentence; for example, let's look at the following sentence.

"I was born in Germany. I speak _____."

Although there isn't necessarily one answer to complete this sentence, as being born in Germany does not predetermine someone to speaking only German, there is a high probability that the missing word is *German*. However, we can only say that because of the context that surrounds the words, and assuming that the neural network was trained on sentences (or phrases) and has a similar structure. Regardless, these types of problems call for a model that can accommodate some sort of memory related to the prior inputs, which brings us to recurrent neural network. Figure 2-5 shows the structure of an RNN.

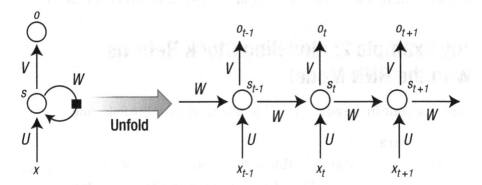

Figure 2-5. *Recurrent neural network*

It is important to examine the structure of the RNN as it relates to resolving the statistical dependency problem. Similar to the prior example, let's walk through some example code in TensorFlow to illustrate the model structure using a toy problem. Similar to the MLP, we will work with a toy problem to create a function that loads and preprocesses our data for the neural network, and then make a function to build our neural network. The following is the beginning of the function:

```
def build_rnn(learning_rate=0.02, epochs=100, state_size=4):
```

The first two arguments should be familiar. They represent the same concepts as in the MLP example. However, we have a new argument called state_size. In a vanilla RNN, the model we are building here, we pass what is called the *hidden state* from a given time step forward. The hidden state is similar to the hidden layer of an MLP in that it is a function of the hidden states at previous time steps. The following defines the hidden state and output as

$$h_t = f\left(W_{xh}x_t + W_{hh}h_{t-1} + b_h\right) \tag{2.9}$$

$$y_t = W_{ho}h_t + b_o \tag{2.10}$$

h_t is the hidden state, W is the weight matrix, b is the bias array, y is the output of the function, and $f(x)$ is the activation function of our choosing.

Toy Example 2: Modeling Stock Returns with the RNN Model

Using the code in the build_rnn() function, observe the following.

```
#Loading data
    x, y = load_data(); scaler = MinMaxScaler(feature_range=(0, 1))
    x, y = scaler.fit_transform(x), scaler.fit_transform(y)
```

```
train_x, train_y = x[0:int(math.floor(len(x)*.67)),  :],
y[0:int(math.floor(len(y)*.67))]

#Creating weights and biases dictionaries
weights = {'input': tf.Variable(tf.random_normal([state_
size+1, state_size])),
    'output': tf.Variable(tf.random_normal([state_size,
    train_y.shape[1]]))}
biases = {'input': tf.Variable(tf.random_normal([1, state_
size])),
    'output': tf.Variable(tf.random_normal([1, train_y.
    shape[1]]))}
```

We begin by loading the training and test data, performing a similar split in the test set such that the first 67% of the complete data set becomes the training set and the remaining 33% becomes the test set. In this instance, we distinguish between two classes, 0 or 1, indicating whether the price went up or down. Moving forward, however, we must refer back to the state size parameter to understand the shape of the matrices we produce, again as TensorFlow variables, for the weight and bias matrices.

To crystallize your understanding of the state size parameter, refer to Figure 2-5, in which the center of the neural network represents a state. We multiply the given input, as well as the previous state, by a weight matrix, and sum all of this with the bias. Similar to the MLP, the weighted sum value forms the input for the activation function.

The output of the activation function forms the hidden state at time step t, whose value becomes part of the weighted sum in Equation 2.10. The value of this matrix application ultimately forms the output for the RNN. We repeat these operations for as many states that we have, which is equal to the number of inputs that we pass through the neural network. When referring back to the image, this is what is meant by the RNN being "unfolded." The state_size in our example is set to 4, meaning that we are inputting four input sequences before we make a prediction.

Let's now walk through the TensorFlow code associated with these operations.

```
#Defining placeholders and variables
    X = tf.placeholder(tf.float32, [batch_size, train_x.shape[1]])
    Y = tf.placeholder(tf.int32, [batch_size, train_y.shape[1]])
    init_state = tf.placeholder(tf.float32, [batch_size, state_
    size])
    input_series = tf.unstack(X, axis=1)
    labels = tf.unstack(Y, axis=1)
    current_state = init_state
    hidden_states = []

#Passing values from one hidden state to the next
for input in input_series: #Evaluating each input within
the series of inputs
    input = tf.reshape(input, [batch_size, 1]) #Reshaping
    input into MxN tensor
    input_state = tf.concat([input, current_state], axis=1)
    #Concatenating input and current state tensors
    _hidden_state = tf.tanh(tf.add(tf.matmul(input_
    state, weights['input']), biases['input'])) #Tanh
    transformation
    hidden_states.append(_hidden_state) #Appending the next
    state
    current_state = _hidden_state #Updating the current state
```

Similar to the MLP model, we need to define place holder variables for both the x and y tensors that our data will pass through. However, a new placeholder will be here, which is the init_state, representing the initial state matrix. Notice that the current state is the init_state placeholder for the first iteration through the next. It also holds the same dimensions and expects the same data type.

Moving forward, we iterate through every input_sequence in the data set, where _hidden_state is the Python definition of formula (see Equation 2.9). Finally, we must come to the output state, given by the following:

```
logits = [tf.add(tf.matmul(state, weights['output']),
biases['output']) for state in hidden_states]
```

The code here is representative of Equation 2.10. However, this will only give us a floating-point decimal, which we need to convert into a label somehow. This brings us to an activation function which will be important to remember for multiclass classification, and therefore for the remainder of this text, the softmax activation function. Subsequently, we define this activation function as the following:

$$S(y_i) = \left(\frac{e^{y^i}}{\sum_{i=1}^{N} e^{y^i}} \right) \tag{2.11}$$

When you look at the formula, we are summing some value over all the possible values. As such, we define this as a probability score. When relating this back to classification, particularly with the RNN, we are outputting the relative probability of an observation being of one class vs another (or others). The label we choose in this instance is the one with the highest relative score, meaning that we choose a given label k because it has the highest probability of being true based on the model's prediction. Equation 2.11 is subsequently represented in the code by the following line:

```
predicted_labels = [tf.nn.softmax(logit) for logit in logits]
#predictions for each logit within the series
```

Being that this is a classification problem, we use a cross entropy-based loss function and for this toy example we will use the gradient descent algorithm, both of which were elaborated upon in the prior section MLPs. Invoking the TensorFlow session also is performed in the same fashion as it would be for the MLP graph (and furthermore for all TensorFlow computational graphs). In a slight derivation from the MLP, we calculate errors at each time step of an unrolled network and sum these errors. This is known as *backpropagation through time* (BPTT), which is utilized specifically because the same weight matrix is used for every time step. As such, the only changing variable besides the input is the hidden state matrix. As such, we can calculate each time step's contribution to the error. We then sum these time step errors to get the error. Mathematically, this is represented by the following equation:

$$\frac{\partial E_3}{\partial W} = \sum_{k=0}^{3} \frac{\partial E_3}{\partial \widehat{y_3}} \frac{\partial y_3}{\partial s_3} \frac{\partial s_3}{\partial s_k} \frac{\partial s_k}{\partial W}$$

This is an application of the chain rule, as described briefly in the section on how we backpropagate the error from the output layer back to the input layer to update the weights with respect to their contribution to the total error. BPTT applies the same logic; instead, we treat the time steps as the layers. However, although RNNs solved many problems of MLPs, they had relative limitations, which you should be aware of.

One of the largest drawbacks of RNNs is that the vanishing gradient problem reappears. However, instead of it being due to having very deep neural network layers, it is caused by trying to evaluate arbitrarily long sequences. The activation function used in RNNs is often the tanh activation function. Mathematically, we define this as follows:

$$\tanh(x) = \frac{e^x - e^{-x}}{e^x + e^{-x}}$$

Figure 2-6 illustrates the activation function.

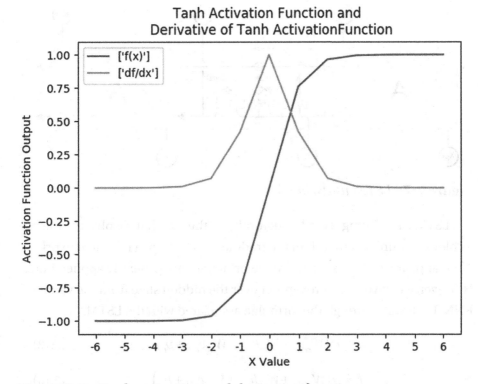

Figure 2-6. *Tanh activation and derivative function*

Similar to the problem with the sigmoid activation function, the derivative of the tanh function can 0, such that when backpropagated over large sequences results in a gradient that is equal to 0. Similar to the MLP, this can cause problems with learning. Depending on the choice of activation function, we also might experience the opposite of the vanishing gradient problem—the exploding gradient. Simply stated, this is the result of the gradients appearing as NaN values. There are couple of solutions for the vanishing gradient function in RNNs. Among them are to try weight regularization via an L1 or L2 norm, or to try different activation functions as we did in the MLP, utilizing functions such as ReLU. However, one of the more straightforward solutions is to use a model devised in the 1990s by Sepp Hochreiter and Jürgen Schmidhuber: the long short-term memory unit, or LSTM. Let's start with what this model looks like, as shown in Figure 2-7.

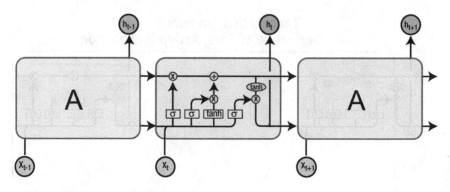

Figure 2-7. *LSTM units/blocks*

LSTMs are distinguished structurally by the fact that we observe them as blocks, or units, rather than the traditional structure a neural network often appears as. That said, the same principles are generally applied here. However, we have an improvement over the hidden state from the vanilla RNN. I will walk through the formulae associated with the LSTM.

$$i_t = \sigma\left(W_{xi}x_t + W_{hi}h_{t-1} + W_{hc}c_{t-1} + b_i\right) \qquad (2.12)$$

$$f_t = \sigma\left(W_{xf}x_t + W_{hf}h_{t-1} + W_{hf}c_{t-1} + b_f\right) \qquad (2.13)$$

$$c_t = f_t \circ c_{t-1} + i_t \circ \tanh\left(W_{xc}x_t + W_{hc}h_{t-1} + b_c\right) \qquad (2.14)$$

$$o_t = \sigma\left(W_{xo}x_t + W_{ho}h_{t-1} + W_{co}c_t + b_o\right) \qquad (2.15)$$

$$h_t = o_t \circ \tanh\left(c_t\right) \qquad (2.16)$$

i_t is the input gate, f_t is the forget gate, c_t is the cell state, o_t is the output gate, h_t is the output vector, σ is the sigmoid activation function, and tanh is the tanh activation function. Both the hidden and cell states are initialized at 0 upon initialization of the algorithm.

The formulae from the LSTM is similar to that of the vanilla RNN, however there is some slight complexity added. Initially, let's draw our attention to the diagram, specifically the LSTM unit in the center, and understand the directional flow as they relate to the formulae.

Preliminarily, let's discuss the notation. Each block, denoted by ▭,
represents a neural network layer, through which we pass through values.
The horizontal lines with arrows represent the vectors and direction in
which the data moves. After it moves through a neural network layer, the
data often is passed to a pointwise operation object, represented by ◯.

Now that I have discussed how to read the diagram, let's dive in deeper.

LSTMs are distinguished by having gates that regulate the information
that passes through individual units, as well as what information passes to
the next unit. Individually, these are the input gate, the output gate, and
the forget gate. In addition to these three gates, an LSTM also contains a
cell, which is an important aspect of the unit.

On the diagram, the cell is represented by the horizontal line, and it
is mathematically represented in Equation 2.14. The cell state is similar
to the hidden state, featured here as well as in the RNN, except there is
discretion as to how much information we pass from one unit to the next.
When looking at the diagram, an input, x_t, is passed through the input
gate. Here the neural network is put through a neural network layer,
with a sigmoid activation function that passes the output to a pointwise
multiplication operator. This operation is combined with the forget gate, f_t,
which is the entirety of Equation 2.14.

Above all, what you should take away from this operation is that its
output is a number between and including 0 and 1. The closer the number
is to 1, information is increasingly passed to the subsequent unit. In
contrast, the closer the number is to 0, information is decreasingly passed
to the subsequent unit.

In Equation 2.13, the forget gate, is what regulates this acceptance of
information, which is represented by c_{t-1}.

Moving to Equation 2.15 and relating it to the diagram, this is the
neural network layer furthest to the right that is passed through another
sigmoid layer, in similar fashion in to the input layer. The output of this
sigmoid activated neural network layer is then multiplied with the tanh
activated cell state vector, in Equation 2.16 Finally, we pass both the

cell state vector and the output vector to the next LSTM unit. While I do not draw out the LSTM in the same fashion as the RNN, I utilize the TensorFlow API's implementation of the LSTM.

Toy Example 3: Modeling Stock Returns with the LSTM Model

As was the case in our prior neural network examples, we must still create TensorFlow placeholders and variables. For this example, the LSTM expects sequences of data, which we facilitate by first creating a three-dimensional X placeholder variables. *To avoid debugging issues when deploying this API with different data sets, you should be careful to read the following instructions carefully.*

```
X = tf.placeholder(tf.float32, (None, None, train_x.shape[1]))
Y = tf.placeholder(tf.float32, (None, train_y.shape[1]))
weights = {'output': tf.Variable(tf.random_normal([n_
hidden, train_y.shape[1]]))}
biases = {'output': tf.Variable(tf.random_normal([train_y.
shape[1]]))}
input_series = tf.reshape(X, [-1, train_x.shape[1]])
input_series = tf.split(input_series, train_x.shape[1], 1)

lstm = rnn.core_rnn_cell.BasicLSTMCell(num_units=n_hidden,
forget_bias=1.0, reuse=None, state_is_tuple=True)
_outputs, states = rnn.static_rnn(lstm, input_series,
dtype=tf.float32)
predictions = tf.add(tf.matmul(_outputs[-1],
weights['output']), biases['output'])
accuracy = tf.reduce_mean(tf.cast(tf.equal(tf.argmax
(tf.nn.softmax(predictions), 1)tf.argmax(Y, 1)), dtype=tf.
float32)),
```

```
error = tf.reduce_mean(tf.nn.softmax_cross_entropy_with_
logits(labels=Y, logits=predictions))
adam_optimizer = tf.train.AdamOptimizer(learning_rate).
minimize(error)
```

When creating a sequence of variables, we start by creating a three-dimensional placeholder named X, which is what we feed our data into. We transform this variable by creating a two-dimensional vector of the observations with the `tf.reshape()`.

Next, we create a tensor object for each of these observations with the `tf.split()` function, which are then stored as a list underneath `input_series`.

Then, we can create an LSTM cell using the `BasicLSTMCell()` function. The `static_rnn()` function accepts any type of RNN cell, so you can utilize other types of RNNs, such as GRUs or vanilla RNNs, and the inputs. Everything else follows the same pattern as the prior examples, in that we create TensorFlow variables to calculate accuracy, the error rate, and the Adam optimizer.

Summary

We have reached the end of our brief, but necessary review of machine learning before we dive deeply into tackling problems using these models on text data. However, it is important for us to review some key concepts:

- **Model choice matters!** Understand the data that you are analyzing. Is the label you are predicting dependent on other prior observed labels, or are these inputs and outputs statistically independent of one another? Failing to inspect these key properties of your data beforehand will waste time and provide you with suboptimal results. Do not skip these steps.

- **Parameter choice matters!** Picking the right model
  for a problem is the first step, but you have to tune this
  model properly to get optimal results. Inspect model
  performance when you alter the number of hidden
  units and epochs. I suggest utilizing algorithms such
  as Adam to tune the learning rate while the network
  is training. Where possible, grid search or use similar
  reactive search methods to find better parameters.

- **Activation functions matter!** Be mindful of how your
  neural network behaves with respect to the vanishing
  gradient problem, particularly if you are working with
  long sequences or have very deep neural networks.

With these concepts in mind, there is one that we did not cover in this
chapter: data preprocessing. It is more appropriate to discuss with the
problems we are facing.

Let's move from this chapter and get into the weeds of natural
language processing with a couple of example problems. In the next
chapter, we walk through a couple of methods for preprocessing text,
discuss their relative advantages and disadvantages, and compare model
performance when using them.

CHAPTER 3

Working with Raw Text

Those who approach NLP with the intention of applying deep learning are most likely immediately confronted with a simple question: How does a machine learning algorithm learn to interpret text data? Similar to the situations in which a feature set may have a categorical feature, we must perform some preprocessing. While the preprocessing we perform in NLP often is more involved than simply converting a categorical feature using label encoding, the principle is the same. We need to find a way to represent individual observations of texts as a row, and encode a static number of features, represented as columns, across all of these observations. As such, feature extraction becomes the most important aspect of text preprocessing.

Thankfully, there has been a considerable amount of work, including ongoing work, to develop preprocessing algorithms of various complexities. This chapter introduces these preprocessing methods, walks through which situations they each work well with, and applies them to example NLP problems that focus on document classification. Let's start by discussing what you should be aware of prior to performing feature extraction from text.

© Taweh Beysolow II 2018
T. Beysolow II, *Applied Natural Language Processing with Python*,
https://doi.org/10.1007/978-1-4842-3733-5_3

Tokenization and Stop Words

When you are working with raw text data, particularly if it uses a web crawler to pull information from a website, for example, you must assume that not all of the text will be useful to extract features from. In fact, it is likely that more noise will be introduced to the data set and make the training of a given machine learning model less effective. As such, I suggest that you perform preliminary steps. Let's walk through these steps using the following sample text.

```
sample_text = "'I am a student from the University of Alabama. I
was born in Ontario, Canada and I am a huge fan of the United
States. I am going to get a degree in Philosophy to improve
my chances of becoming a Philosophy professor. I have been
working towards this goal for 4 years. I am currently enrolled
in a PhD program. It is very difficult, but I am confident that
it will be a good decision"'
```

When the `sample_text` variable prints, there is the following output:

```
'I am a student from the University of Alabama. I
was born in Ontario, Canada and I am a huge fan of the United
States. I am going to get a degree in Philosophy to improve my
chances of becoming a Philosophy professor. I have been working
towards this goal for 4 years. I am currently enrolled in a PhD
program. It is very difficult, but I am confident that it will
be a good decision'
```

You should observe that the computer reads bodies of text, even if punctuated, as single string objects. Because of this, we need to find a way to separate this single body of text so that the computer evaluates each word as an individual string object. This brings us to the concept of *word tokenization*, which is simply the process of separating a single string object, usually a body of text of varying length, into individual tokens that represent words or characters that we would like to evaluate further. Although you can find ways to implement this from scratch, for brevity's sake, I suggest that you utilize the Natural Language Toolkit (NLTK) module.

NLTK allows you to use some of the more basic NLP functionalities, as well as pretrained models for different tasks. It is my goal to allow you to train your own models, so we will not be working with any of the pretrained models in NLTK. However, you should read through the NLTK module documentation to become familiar with certain functions and algorithms that expedite text preprocessing. Relating back to our example, let's tokenize the sample data via the following code:

```
from nltk.tokenize import word_tokenize, sent_tokenize
sample_word_tokens = word_tokenize(sample_text)
sample_sent_tokens = sent_tokenize(sample_text)
```

When you print the sample_word_tokens variable, you should observe the following:

```
['I', 'am', 'a', 'student', 'from', 'the', 'University', 'of',
'Alabama', '.', 'I', 'was', 'born', 'in', 'Ontario', ',',
'Canada', 'and', 'I', 'am', 'a', 'huge', 'fan', 'of', 'the',
'United', 'States', '.', 'I', 'am', 'going', 'to', 'get', 'a',
'degree', 'in', 'Philosophy', 'to', 'improve', 'my', 'chances',
'of', 'becoming', 'a', 'Philosophy', 'professor', '.', 'I',
'have', 'been', 'working', 'towards', 'this', 'goal', 'for',
'4', 'years', '.', 'I', 'am', 'currently', 'enrolled', 'in',
'a', 'PhD', 'program', '.', 'It', 'is', 'very', 'difficult',
',', 'but', 'I', 'am', 'confident', 'that', 'it', 'will', 'be',
'a', 'good', 'decision']
```

You will also observe that we have defined another tokenized object, sample_sent_tokens. The difference between word_tokenize() and sent_tokenize() is simply that the latter tokenizes text by sentence delimiters. This is observed in the following output:

```
['I am a student from the University of Alabama.', 'I \nwas
born in Ontario, Canada and I am a huge fan of the United
States.', 'I am going to get a degree in Philosophy to improve
my chances of \nbecoming a Philosophy professor.', 'I have
been working towards this goal\nfor 4 years.', 'I am currently
enrolled in a PhD program.', 'It is very difficult, \nbut I am
confident that it will be a good decision']
```

Now we have individual tokens that we can preprocess! From this step forward, we can clean out some of the junk text that we would not want to extract features from. Typically, the first thing we want to get rid of are *stop words*, which are usually defined as very common words in a given language. Most often, lists of stop words that we build or utilize in software packages include *function words*, which are words that express a grammatical relationship (rather than having an intrinsic meaning). Examples of function words include *the, and, for,* and *of*.

In this example, we use the list of stop words from the NLTK package.

```
[u'i', u'me', u'my', u'myself', u'we', u'our', u'ours',
u'ourselves', u'you', u"you're", u"you've", u"you'll",
u"you'd", u'your', u'yours', u'yourself', u'yourselves',
u'he', u'him', u'his', u'himself', u'she', u"she's", u'her',
u'hers', u'herself', u'it', u"it's", u'its', u'itself',
u'they', u'them', u'their', u'theirs', u'themselves', u'what',
u'which', u'who', u'whom', u'this', u'that', u"that'll",
u'these', u'those', u'am', u'is', u'are', u'was', u'were',
```

```
u'be', u'been', u'being', u'have', u'has', u'had', u'having',
u'do', u'does', u'did', u'doing', u'a', u'an', u'the', u'and',
u'but', u'if', u'or', u'because', u'as', u'until', u'while',
u'of', u'at', u'by', u'for', u'with', u'about', u'against',
u'between', u'into', u'through', u'during', u'before',
u'after', u'above', u'below', u'to', u'from', u'up', u'down',
u'in', u'out', u'on', u'off', u'over', u'under', u'again',
u'further', u'then', u'once', u'here', u'there', u'when',
u'where', u'why', u'how', u'all', u'any', u'both', u'each',
u'few', u'more', u'most', u'other', u'some', u'such', u'no',
u'nor', u'not', u'only', u'own', u'same', u'so', u'than',
u'too', u'very', u's', u't', u'can', u'will', u'just', u'don',
u"don't", u'should', u"should've", u'now', u'd', u'll',
u'm', u'o', u're', u've', u'y', u'ain', u'aren', u"aren't",
u'couldn', u"couldn't", u'didn', u"didn't", u'doesn',
u"doesn't", u'hadn', u"hadn't", u'hasn', u"hasn't", u'haven',
u"haven't", u'isn', u"isn't", u'ma', u'mightn', u"mightn't",
u'mustn', u"mustn't", u'needn', u"needn't", u'shan', u"shan't",
u'shouldn', u"shouldn't", u'wasn', u"wasn't", u'weren',
u"weren't", u'won', u"won't", u'wouldn', u"wouldn't"]
```

All of these words are lowercase by default. You should be aware that string objects must exactly match to return a true Boolean variable when comparing two individual strings. To put this more plainly, if we were to execute the code *"you"* == *"YOU"*, the Python interpreter returns false. The specific instance in which this affects our example can be observed by executing the mistake() and advised_preprocessing() functions, respectively. Observe the following outputs:

```
['I', 'student', 'University', 'Alabama', '.', 'I', 'born',
'Ontario', ',', 'Canada', 'I', 'huge', 'fan', 'United',
'States', '.', 'I', 'going', 'get', 'degree', 'Philosophy',
'improve', 'chances', 'becoming', 'Philosophy', 'professor',
'.', 'I', 'working', 'towards', 'goal', '4', 'years', '.',
'I', 'currently', 'enrolled', 'PhD', 'program', '.', 'It',
'difficult', ',', 'I', 'confident', 'good', 'decision']
```

```
['student', 'University', 'Alabama', '.', 'born', 'Ontario',
',', 'Canada', 'huge', 'fan', 'United', 'States', '.',
'going', 'get', 'degree', 'Philosophy', 'improve', 'chances',
'becoming', 'Philosophy', 'professor', '.', 'working',
'towards', 'goal', '4', 'years', '.', 'currently', 'enrolled',
'PhD', 'program', '.', 'difficult', ',', 'confident', 'good',
'decision']
```

As you can see, the mistake() function does not catch the uppercase "I" characters, meaning that there are several stop words still in the text. This is solved by uppercasing all the stop words and then evaluating whether each uppercase word in the sample text was in the stop_words list. This is exemplified with the following two lines of code:

```
stop_words = [word.upper() for word in stopwords.
words('english')]
word_tokens = [word for word in sample_word_tokens if word.
upper() not in stop_words]
```

Although embedded methods in feature extraction algorithms likely account for this case, you should be aware that strings must match exactly, and you must account for this when preprocessing manually.

That said, there is junk data that you should be aware of—specifically, the grammatical characters. You will be relieved to hear that the word_tokenize() function also categorizes colons and semicolons as individual word tokens, but you still have to get rid of them. Thankfully, NLTK contains another tokenizer worth knowing about, which is defined and utilized in the following code:

```
from nltk.tokenize import RegexpTokenizer
tokenizer = RegexpTokenizer(r'\w+')
sample_word_tokens = tokenizer.tokenize(str(sample_word_
tokens))
sample_word_tokens = [word.lower() for word in sample_word_
tokens]
```

When we print the sample_word_tokens variable, we get the following output:

```
['student', 'university', 'alabama', 'born', 'ontario',
'canada', 'huge', 'fan', 'united', 'states', 'going', 'get',
'degree', 'philosophy', 'improve', 'chances', 'becoming',
'philosophy', 'professor', 'working', 'towards', 'goal',
'4', 'years', 'currently', 'enrolled', 'phd', 'program',
'difficult', 'confident', 'good', 'decision']
```

In the course of this example, we have reached the final step! We have removed all the standard stop words, as well as all grammatical tokens. This is an example of a document that is ready for feature extraction, whereupon some additional preprocessing may occur.

Next, I'll discuss some of the various feature extraction algorithms. And let's work on denser sample data alongside a preprocessed example paragraph.

The Bag-of-Words Model (BoW)

A BoW model is one of the more simplistic feature extraction algorithms that you will come across. The name "bag-of-words" comes from the algorithm simply seeking to know the number of times a given word is present within a body of text. The order or context of the words is not analyzed here. Similarly, if we have a bag filled with six pencils, eight pens, and four notebooks, the algorithm merely cares about recording the number of each of these objects, not the order in which they are found, or their orientation.

Here, I have defined a sample bag-of-words function.

```
def bag_of_words(text):
    _bag_of_words = [collections.Counter(re.findall(r'\w+',
    word)) for word in text]
    bag_of_words = sum(_bag_of_words, collections.Counter())
    return bag_of_words

sample_word_tokens_bow = bag_of_words(text=sample_word_tokens)
print(sample_word_tokens_bow)
```

When we execute the preceding code, we get the following output:

```
Counter({'philosophy': 2, 'program': 1, 'chances': 1, 'years': 1,
'states': 1, 'born': 1, 'towards': 1, 'canada': 1, 'huge': 1,
'united': 1, 'goal': 1, 'working': 1, 'decision': 1,
'currently': 1, 'confident': 1, 'going': 1, '4': 1,
'difficult': 1, 'good': 1, 'degree': 1, 'get': 1, 'becoming': 1,
'phd': 1, 'ontario': 1, 'fan': 1, 'student': 1, 'improve': 1,
'professor': 1, 'enrolled': 1, 'alabama': 1, 'university': 1})
```

This is an example of a BoW model when presented as a dictionary. Obviously, this is not a suitable input format for a machine learning algorithm. This brings me to discuss the myriad of text preprocessing functions available in the scikit-learn library, which is a Python library that all data scientists and machine learning engineers should be familiar with. For those who are new to it, this library provides implementations of machine learning algorithms, as well as several data preprocessing algorithms. Although we won't walk through much of this package, the text preprocessing functions are extremely useful.

CountVectorizer

Let's start by walking through the BoW equivalent—CountVectorizer, an implementation of bag-of-words in which we code text data as a representation of features/words. The values of each of these features represent the occurrence counts of words across all documents. If you recall, we defined a sample_sent_tokens variable, which we will analyze. We define a bow_sklearn() function beneath where we preprocess our data. The function is defined as follows:

```
from sklearn.feature_extraction.text import CountVectorizer
def bow_sklearn(text=sample_sent_tokens):
    c = CountVectorizer(stop_words='english',
    token_pattern=r'\w+')
    converted_data = c.fit_transform(text).todense()
    print(converted_data.shape)
    return converted_data, c.get_feature_names()
```

To provide context, in this example, we are assuming that each sentence is an individual document, and we are creating a feature set in which each feature is an individual token. When we instantiate CountVectorizer(), we set two parameters: stop_words, and token_pattern. These two arguments are the embedded methods in the feature extraction that remove stop words and grammatical tokens. The fit_transform() attribute expects to receive a list, an array, or a similar object of iterable string objects. We assign the bow_data and feature_names variables to the data that the `bow_sklearn()` returns, respectively. Our converted data set is a 6 × 50 matrix, which means that we have six sentences, all of which have 50 features. Observe our data set and feature names, respectively, in the following outputs:

```
[[0 1 0 0 0 0 0 0 0 0 0 0 0 0 0 0 0 0 0 0 0 0 0 0 0 1 0 1 0 0]
 [0 0 1 1 0 0 0 0 0 0 0 1 0 0 0 1 0 1 0 0 0 0 1 0 1 0 0 0]
 [0 0 0 0 1 0 0 0 1 0 0 0 0 1 0 0 1 0 0 1 0 0 2 1 0 0 0 0 0 0]
 [1 0 0 0 0 0 0 0 0 0 0 0 0 1 0 0 0 0 0 0 0 0 0 0 0 0 0 1 1]
 [0 0 0 0 0 0 1 0 0 0 1 0 0 0 0 0 0 0 0 1 0 0 1 0 0 0 0 0 0]
 [0 0 0 0 0 1 0 1 0 1 0 0 0 0 1 0 0 0 0 0 0 0 0 0 0 0 0 0 0]]
```

```
[u'4', u'alabama', u'born', u'canada', u'chances',
u'confident', u'currently', u'decision', u'degree',
u'difficult', u'enrolled', u'fan', u'goal', u'going', u'good',
u'huge', u'improve', u'ontario', u'phd', u'philosophy',
u'professor', u'program', u'states', u'student', u'united',
u'university', u'working', u'years']
```

To extrapolate this example to a larger number of documents, and ostensibly larger vocabulary sizes, our matrices for preprocessed text data tends to have a large number of features, sometimes well over 1000. How to evaluate these features effectively is the machine learning challenge we seek to solve. You typically want to use the bag-of-words feature extraction technique for document classification. Why is this the case? We assume that documents of certain classifications contain certain words. For example, we expect a document referencing political science to perhaps feature jargon such as *dialectical materialism* or *free market capitalism*; whereas a document that is referring to classical music will have terms such as *crescendo*, *diminuendo*, and so forth. In these instances of document classification, the location of the word itself is not terribly important. It's important to know what portion of the vocabulary is present in one class of document vs. another.

Next, let's look at our first example problem in the code in the text_classifiction_demo.py file.

Example Problem 1: Spam Detection

Spam detection is a relatively common task in that most people have an inbox (email, social media instant messenger account, or similar entity) targeted by advertisers or malicious actors. Being able to block unwanted advertisements or malicious files is an important task. Because of this, we are interested in pursuing a machine learning approach to spam detection. Let's begin by describing the data set before digging into the problem.

This data set was downloaded from the UCI Machine Learning Repository, specifically the Text Data section. Our data set consists of 5574 observations—all SMS messages. We observe from our data set that most of the messages are not terribly long. Figure 3-1 is a histogram of our entire data set.

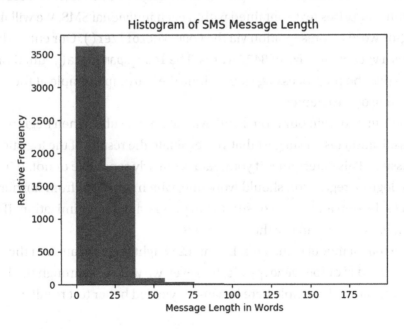

Figure 3-1. *SMS message length histogram*

Something else we should be mindful of is the distribution between the class labels, which tends to be heavily skewed. In this data set, 4825 observations are marked as "ham" (being not spam), and 747 are marked as "spam". You must be vigilant in evaluating your machine learning solutions to ensure that they do not overfit the training data, and then fail miserably on test data.

Let's briefly do some additional data set discovery before we move on to tackling the problem directly. When we look at the header of our data set, we observe the following:

```
0    ham  Go until jurong point, crazy.. Available only ...
1    ham                        Ok lar... Joking wif u oni...
2   spam  Free entry in 2 a wkly comp to win FA Cup fina...
3    ham  U dun say so early hor... U c already then say...
4    ham  Nah I don't think he goes to usf, he lives aro...
```

The first column is our categorical label/response variable. The second column comprises text contained within each individual SMS. We will use a bag-of-words representation via the CountVectorizer(). Our entire data set has a vocabulary size of 8477 words. The load_spam_data() function shows that the preprocessing steps mimic the warmup example at the beginning of the chapter.

Let's fit and train our model and evaluate the results. When beginning a classification task, I suggest that you evaluate the results of the logistic regression. This determines if your data is linearly separable or not. If it is, the logistic regression should work fine, which saves you from further model selection and time-consuming hyper-parameter optimization. If it does fail, then you can use those methods.

We train a model using both L1 and L2 weight regularization in the text_classifiction_demo.py file; however, we will walk through the L1 norm regularized example here because it yielded better test results:

```
#Fitting training algorithm
l = LogisticRegression(penalty='l1')
accuracy_scores, auc_scores = [], []
```

Those of you that are not familiar with logistic regression you should learn about elsewhere; however, I will discuss the L1-regularized logistic regression briefly. L1 norm regularization in linear models is standard for LASSO (least absolute shrinkage selection operator), where during the learning process, the L1 norm can theoretically force some regression coefficients to 0. In contrast, the L2 norm, often seen in ridge regression, can force some regression coefficients during the learning process to numbers close to 0. The difference between this is that coefficients that are 0 effectively perform feature selection on our feature set by eliminating them. Mathematically, we represent this regularization via Equation 3.1.

$$\min \sum_{i=1}^{M} -\log p(y^i | x^I; \theta) + \beta \|\theta\|_1 \qquad (3.1)$$

We will evaluate the distribution of test scores over several trials. scikit-learn algorithms' `fit()` method trains the algorithm of a given data set. As such, all the iterations that optimize the parameters are performed. To see logging information in the training process, set the `verbose` parameter to 1.

Let's look at the code that will collect the distribution of both accuracy and AUC scores.

```
for i in range(trials):
    if i%10 == 0 and i > 0:
        print('Trial ' + str(i) + ' out of 100 completed')
    l.fit(train_x, train_y)
    predicted_y_values = l.predict(train_x)
    accuracy_scores.append(accuracy_score(train_y, predicted_y_
    values))
```

```
fpr, tpr = roc_curve(train_y, predicted_y_values)[0],
roc_curve(train_y, predicted_y_values)[1]
auc_scores.append(auc(fpr, tpr))
```

scikit-learn performs cross-validation so long as you define a random seed utilizing the `np.random.seed()` function, which we do near the beginning of the file. During each trial, we are fitting the data set to the algorithm, predicting the accuracy and AUC score, and appending them to a list that we defined. When we evaluate our results from training, we observe the following:

```
Summary Statistics (AUC):
        Min       Max       Mean      SDev      Range
0   0.965348  0.968378  0.967126  0.000882  0.00303

Summary Statistics (Accuracy Scores):
        Min       Max       Mean      SDev      Range
0   0.990356  0.99116   0.990828  0.000234  0.000804

Test Model Accuracy: 0.9744426318651441
Test True Positive Rate: 0.8412698412698413
Test False Positive Rate: 0.004410838059231254

[[1580    7]
 [  40  212]]
```

Fortunately, we see that logistic regression performs excellently on this problem. We have excellent accuracy and AUC scores, with very little variance from one trial to the next. Let's evaluate the AUC score, as shown in Figure 3-2.

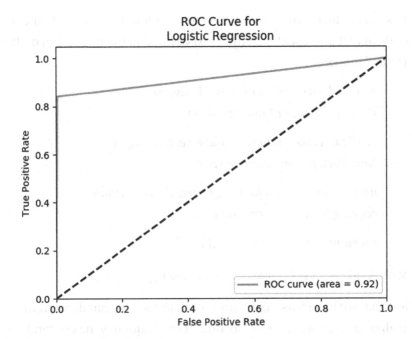

Figure 3-2. *Test set ROC curve*

Our test AUC score is 0.92. This algorithm would be deployable in an application to test for spam results. In the course of solution discovery, I suggest that you use this model rather than others. Although you are encouraged to find other methods, I observed that the gradient-boosted classification tree and random forests performed considerably worse, with AUC scores of roughly 0.72. Let's discuss a more sophisticated term frequency scheme.

Term Frequency Inverse Document Frequency

Term frequency–inverse document frequency (TFIDF) is based on BoW, but provides more detail than simply taking term frequency, as was done in the prior example. TFIDF yields a value that shows how important a given word is by not only looking at term frequency, but also analyzing how many times the word appears across all documents. The first portion, term frequency, is relatively straightforward.

57

Let's look at an example to see how to calculate TFIDF. We define a new body of text and use the sample text defined at the beginning of the chapter, as follows:

> text = "'I was a student at the University of
> Pennsylvania, but now work on
>
> Wall Street as a Lawyer. I have been living in
> New York for roughly five years
>
> now, however I am looking forward to eventually
> retiring to Texas once I have
>
> saved up enough money to do so.'"

```
document_list = list([sample_text, text])
```

Now that we have a list of documents, let's look at exactly what the TFIDF algorithm does. The first portion, term frequency, has several variants, but we will focus on the standard raw count scheme. We simply sum the terms across all documents. The term frequency is equivalent to Equation 3.2.

$$\frac{f_{t,d}}{\sum_{t' \in d} f_{t',d}} \tag{3.2}$$

$f_{t,d}$ is equal to the frequency of the term across all documents. $f_{t',d}$ is equal to the frequency of that same term but within each individual document. In our code, we document these steps in the tf_idf_example() function, as follows:

```
def tf_idf_example(textblobs=[text, text2]):

def term_frequency(word, textblob): (1)
```

```
return textblob.words.count(word)/float(len(textblob.words))

def document_counter(word, text):
return sum(1 for blob in text if word in blob)

def idf(word, text): (2)
return np.log(len(text) /1 + float(document_counter(word,
text)))

def tf_idf(word, blob, text):
return term_frequency(word, blob) * idf(word, text)

output = list()
for i, blob in enumerate(textblobs):
output.append({word: tf_idf(word, blob, textblobs) for word in
blob.words})
print(output)
```

Thanks to the TextBlob package, we are able to fairly quickly re-create the TFIDF toy implementation. I will address each of the functions within the tf_idf_example() function. You are aware of term frequency, so I can discuss inverse document frequency. We define inverse document frequency as a measure of how frequently a word appears across the entire corpus. Mathematically, this relationship is expressed in Equation 3.3.

$$\mathrm{idf}(t,D) = \log\frac{N}{|\{d \in D : t \in d\}|}$$
(3.3)

This equation calculates the log of the total number of documents in our corpus, divided by all the documents in which the term that we are evaluating appears. In our code, we calculate this with the function (2). Now, we are ready to proceed to the final step of the algorithm, which is multiplying the term frequency by the inverse document frequency, as shown in the preceding code. We then yield the following output:

[{'up': '0.027725887222397813', 'money':
'0.021972245773362195', 'am': '0.027725887222397813', 'years':
'0.027725887222397813', 'as': '0.027725887222397813', 'at':
'0.027725887222397813', 'have': '0.055451774444795626',
'in': '0.027725887222397813', 'New': '0.021972245773362195',
'saved': '0.021972245773362195', 'Texas':
'0.021972245773362195', 'living': '0.021972245773362195',
'for': '0.027725887222397813', 'to': '0.08317766166719343',
'retiring': '0.027725887222397813', 'been':
'0.021972245773362195', 'looking': '0.021972245773362195',
'Pennsylvania': '0.021972245773362195', 'enough':
'0.021972245773362195', 'York': '0.021972245773362195',
'forward': '0.027725887222397813', 'was':
'0.027725887222397813', 'eventually': '0.021972245773362195',
'do': '0.027725887222397813', 'I': '0.11090354888959125',
'University': '0.027725887222397813', 'however':
'0.027725887222397813', 'but': '0.021972245773362195', 'five':
'0.021972245773362195', 'student': '0.021972245773362195',
'now': '0.04394449154672439', 'a': '0.055451774444795626',
'on': '0.027725887222397813', 'Wall': '0.021972245773362195',
'of': '0.027725887222397813', 'work': '0.021972245773362195',
'roughly': '0.021972245773362195', 'Street':
'0.021972245773362195', 'so': '0.021972245773362195', 'Lawyer':
'0.021972245773362195', 'the': '0.027725887222397813', 'once':
'0.021972245773362195'}, {'and': '0.0207285337484549', 'is':
'0.0207285337484549', 'each': '0.0207285337484549', 'am':
'0.026156497379620575', 'years': '0.026156497379620575',
'have': '0.05231299475924115', 'in': '0.026156497379620575',
'children': '0.0414570674969098', 'considering':
'0.0207285337484549', 'retirement': '0.0207285337484549',
'doctor': '0.0207285337484549', 'retiring':

'0.026156497379620575', 'two': '0.0207285337484549', 'long':
'0.0207285337484549', 'next': '0.0207285337484549', 'to':
'0.05231299475924115', 'forward': '0.026156497379620575',
'was': '0.026156497379620575', 'couple': '0.0207285337484549',
'more': '0.0207285337484549', 'ago': '0.0207285337484549',
'them': '0.0207285337484549', 'that': '0.0207285337484549',
'I': '0.1046259895184823', 'University':
'0.026156497379620575', 'who': '0.0414570674969098', 'however':
'0.026156497379620575', 'quite': '0.0207285337484549',
'me': '0.0207285337484549', 'Yale': '0.0207285337484549',
'with': '0.0207285337484549', 'the': '0.05231299475924115',
'a': '0.07846949213886173', 'both': '0.0207285337484549',
'look': '0.026156497379620575', 'of': '0.026156497379620575',
'grandfather': '0.0207285337484549', 'spending':
'0.0207285337484549', 'three': '0.0207285337484549', 'time':
'0.0414570674969098', 'making': '0.0207285337484549', 'went':
'0.0207285337484549'}]

This brings us to the end of our toy example using TFIDF. Before we jump into the example, let's review how we would utilize this example in scikit-learn, such that we can input this data into a machine learning algorithm. Similar to CountVectorizer(), scikit-learn has provided a TfidfVectorizer() method that comes in handy. The following shows its utilization. I will dive into a deeper use of its preprocessing methods later.

```
def tf_idf_sklearn(document=document_list):
    t = TfidfVectorizer(stop_words='english',
    token_pattern=r'\w+')
    x = t.fit_transform(document_list).todense()
    print(x)
```

When we execute the function, it yields the following result:

```
[[0.          0.          0.          0.          0.          0.24235766
  0.17243947 0.          0.24235766 0.24235766 0.          0.
  0.24235766 0.          0.24235766 0.24235766 0.24235766 0.
  0.          0.17243947 0.24235766 0.24235766 0.          0.24235766
  0.24235766 0.24235766 0.          0.17243947 0.24235766 0.
  0.24235766 0.          0.17243947 0.24235766]
 [0.20840129 0.41680258 0.20840129 0.20840129 0.20840129 0.
  0.14827924 0.20840129 0.          0.          0.20840129 0.20840129
  0.          0.20840129 0.          0.          0.          0.20840129
  0.20840129 0.14827924 0.          0.          0.20840129 0.
  0.          0.          0.41680258 0.14827924 0.          0.20840129
  0.          0.20840129 0.14827924 0.          ]]
```

This function yields a 2 × 44 matrix, and it is ready for input into a machine learning algorithm for evaluation.

Now let's work through another example problem using TFIDF as our feature extractor while utilizing another machine learning algorithm as we did for the BoW feature extraction.

Example Problem 2: Classifying Movie Reviews

We obtained the following IMDB movie review data set from http://www.cs.cornell.edu/people/pabo/movie-review-data/.

We are going to work with the raw text directly, rather than using preprocessed text data sets often provided via various machine learning packages.

Let's take a snapshot of the data.

```
tristar / 1 : 30 / 1997 / r ( language , violence , dennis
rodman ) cast : jean-claude van damme ; mickey rourke ; dennis
rodman ; natacha lindinger ; paul freeman director : tsui hark
```

screenplay : dan jakoby ; paul mones ripe with explosions ,
mass death and really weird hairdos , tsui hark's " double
team " must be the result of a tipsy hollywood power lunch
that decided jean-claude van damme needs another notch on his
bad movie-bedpost and nba superstar dennis rodman should have
an acting career . actually , in " double team , " neither's
performance is all that bad . i've always been the one critic
to defend van damme -- he possesses a high charisma level that
some genre stars (namely steven seagal) never aim for ; it's
just that he's never made a movie so exuberantly witty since
1994's " timecop . " and rodman . . . well , he's pretty much
rodman . he's extremely colorful , and therefore he pretty much
fits his role to a t , even if the role is that of an ex-cia

As you can see, this data is filled with lots of grammatical noise that we
will need to remove, but is also rich in descriptive text. We will opt to use
the TfidfVectorizer() method on this data.

First, I would like to direct you to two functions at the beginning of
the file:

```
def remove_non_ascii(text):
    return "".join([word for word in text if ord(word) < 128])
```

Notice that we are using the native Python function ord(). This
function expects a string, and it returns either the Unicode point for
Unicode objects or the value of the byte. If the ord() function returns
an integer less than 128, this poses no problem for our preprocesser
and therefore we keep the string in question; otherwise, we remove
the character. We end this step by joining all the remaining words back
together with the ".join() function. The reasoning for preprocessing
during data preparation is that our text preprocessor expects Unicode
objects when being fed to it. When we are capturing raw text data,
particularly if it is from an HTML page, many of the string objects

loaded before preprocessing and removal of stop words will not be Unicode-compatible.

Let's look at the function that loads our data.

```
def load_data():
    negative_review_strings = os.listdir('/Users/tawehbeysolow/
    Downloads/review_data/tokens/neg')
    positive_review_strings = os.listdir('/Users/tawehbeysolow/
    Downloads/review_data/tokens/pos')
    negative_reviews, positive_reviews = [], []
```

We start by loading the file names of all the .txt files to be processed. To do this, we use the os.listdir() function. I suggest you use this function when building similar applications that require preprocessing a large number of files.

Next, we load our files with the open() function, and then apply the remove_non_ascii() function, as follows:

```
for positive_review in positive_review_strings:
    with open('/Users/tawehbeysolow/Downloads/review_data/
    tokens/pos/'+str(positive_review), 'r') as positive_file:
        positive_reviews.append(remove_non_ascii(positive_
        file.read()))

for negative_review in negative_review_strings:
    with open('/Users/tawehbeysolow/Downloads/review_data/
    tokens/neg/'+str(negative_review), 'r') as negative_file:
        negative_reviews.append(remove_non_ascii(negative_
        file.read()))
```

With our initial preprocessing done, we end by concatenating both the positive and negative reviews, in addition to the respective vectors that contain their labels. Now, we can get to the meat and potatoes of this machine learning problem, starting with the train_logistic_model()

function. In a similar fashion, we use logistic regression as the baseline for the problem. Although most of the following functions are similar in structure to Example Problem 1, let's look at the beginning of this function to analyze what we have changed.

```
#Load and preprocess text data
x, y = load_data()
t = TfidfVectorizer(min_df=10, max_df=300, stop_
words='english', token_pattern=r'\w+')
x = t.fit_transform(x).todense()
```

We are utilizing two new arguments: min_df corresponds to the minimum document frequency to retain a word, and max_df refers to the maximum amount of documents that a word can appear in before it is omitted from the sparse matrix that we create. When increasing the maximum and minimum document frequencies, I noticed that the L1 penalty model performed better than the L2 penalty model. I would posit that this is likely due to the fact that as we increase the min_df parameter, we are creating a considerably sparser matrix than if we had a denser matrix. You should keep this in mind so as to not overselect features if they performed any feature selection on their matrices beforehand.

Let's evaluate the results of the logistic regression, as shown in the following output (also see Figures 3-3 and 3-4).

```
Summary Statistics from Training Set (AUC):
       Mean        Max  Range       Mean  SDev
0  0.723874  0.723874    0.0  0.723874   0.0
Summary Statistics from Training Set (Accuracy):
       Mean        Max  Range       Mean  SDev
0  0.726788  0.726788    0.0  0.726788   0.0
Training Data Confusion Matrix:
[[272 186]
 [ 70 409]]
```

Summary Statistics from Test Set (AUC):
```
        Mean        Max  Range        Mean  SDev
0   0.723874   0.723874     0.0   0.723874   0.0
```
Summary Statistics from Test Set (Accuracy):
```
        Mean        Max  Range        Mean  SDev
0   0.726788   0.726788     0.0   0.726788   0.0
```
Test Data Confusion Matrix:
```
[[272 186]
 [ 70 409]]
```

Summary Statistics from Training Set (AUC):
```
        Mean        Max  Range        Mean  SDev
0   0.981824   0.981824     0.0   0.981824   0.0
```
Summary Statistics from Training Set (Accuracy):
```
        Mean        Max  Range        Mean  SDev
0   0.981857   0.981857     0.0   0.981857   0.0
```
Training Data Confusion Matrix:
```
[[449   9]
 [  8 471]]
```

Summary Statistics from Test Set (AUC):
```
        Mean        Max  Range        Mean  SDev
0   0.981824   0.981824     0.0   0.981824   0.0
```
Summary Statistics from Test Set (Accuracy):
```
        Mean        Max  Range        Mean  SDev
0   0.981857   0.981857     0.0   0.981857   0.0
```

Test Data Confusion Matrix:
```
[[449   9]
 [  8 471]]
```

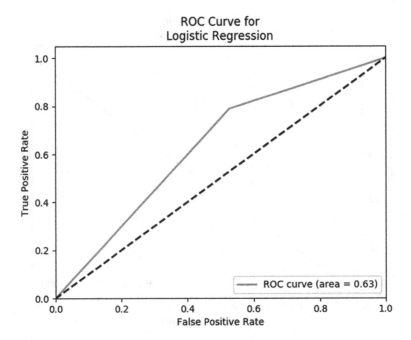

Figure 3-3. *L1 logistic regression test set ROC curve*

Both in training and performance, logistic regression performs considerably better when utilizing the L2 weight regularization method, given the parameters we used for the `TfidfVectorizer()` feature extraction algorithm.

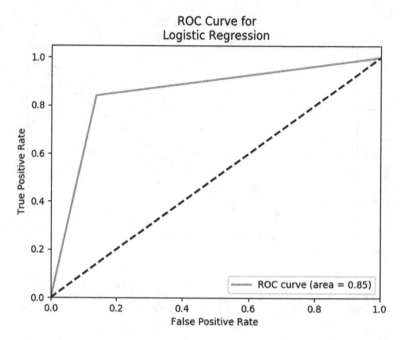

Figure 3-4. *L2 logistic regression test set ROC curve*

I created multiple solutions to evaluate: a random forest classifier, a naïve Bayes classifier, and a multilayer perceptron. We begin with a general overview of all of our methods and their respective orientations.

Starting with the multilayer perceptron in the `mlp_movie_classification_model.py` file, notice that much of the neural network is the same as the example in Chapter 2, with the exception of an extra hidden layer. That said, I would like to direct your attention to lines 92 through 94.

```
regularization = tf.contrib.layers.l2_regularizer(scale=0.0005,
scope=None)
regularization_penalty = tf.contrib.layers.apply_
regularization(regularization, weights.values())
cross_entropy = cross_entropy + regularization_penalty
```

In these lines, we are performing weight regularization, as discussed earlier in this chapter with the logistic regression L2 and L1 loss parameters. Those of you who wish to apply this in TensorFlow can rest assured that these are the only modifications needed to add a weight penalty to your neural network. While developing this solution, I tried weight regularization utilizing L1 and L2 loss penalties, and I experimented with dropout. Weight regularization is the process of limiting the scope to which the weights can grow when utilizing different vector norms. The two most referenced norms for weight regularization are L1 and L2 norms. The following are their respective equations, which are also illustrated in Figure 3-5.

$$L_1 = \|\mathbf{v}\|_1 = \sum_{i=1}^{N} |v_i|^1$$

$$L_2 = \|\mathbf{v}\|_2 = \sqrt{\sum_{i=1}^{N} |v_i|^2}$$

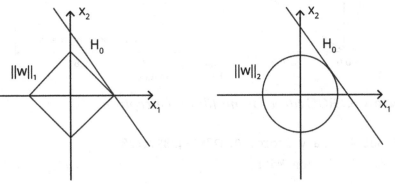

Figure 3-5. *L1 and L2 norm visualization*

When initially utilizing both one and two hidden layer(s), I noticed that both the test and training performance were considerably worse with dropout, even using dropout percentages as low as 0.05. As such, I cannot suggest that you utilize dropout for this problem. As for weight regularization, additional parameter selection is not advisable; however, I found negligible differences with L1 vs. L2 regularization. The confusion matrix and the ROC curve are shown in Figure 3-6.

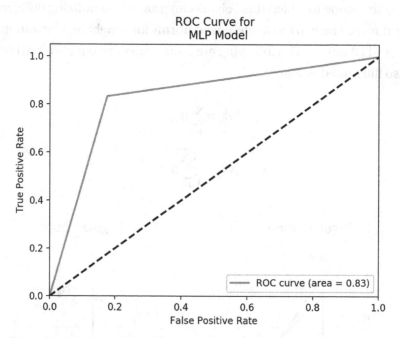

Figure 3-6. *ROC curve for multilayer perceptron*

```
Test Set Accuracy Score: 0.8285714285714286
Test Set Confusion Matrix:
[[122  26]
 [ 22 110]]
```

Let's analyze the choice of parameters for the random forest and naïve Bayes classifiers. We kept our trees relatively short at a max_depth of ten splits. As for the naïve Bayes classifier, the only parameter we chose is alpha, which we set to 0.005. Let's evaluate Figures 3-6 and 3-7 for the results of the model.

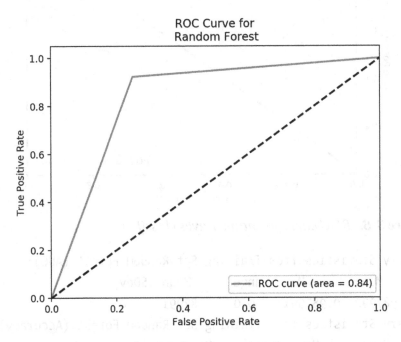

Figure 3-7. *ROC curve for random forest*

The Figure 3-8 shows the result of a naïve Bayes classifier.

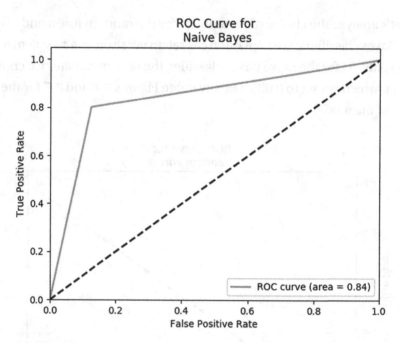

Figure 3-8. *ROC curve for naïve Bayes classifier*

```
Summary Statistics from Training Set Random Forest (AUC):
        Mean        Max   Range       Mean   SDev
0  0.987991  0.987991    0.0   0.987991   0.0
Summary Statistics from Training Set Random Forest (Accuracy):
        Mean       Max   Range       Mean   SDev
0  0.98826   0.98826    0.0   0.98826    0.0
Training Data Confusion Matrix (Random Forest):
[[447  11]
 [  0 479]]
Summary Statistics from Training Set Naive Bayes (AUC):
        Mean        Max   Range       Mean           SDev
0  0.965362  0.965362    0.0   0.965362   2.220446e-16
```

```
Summary Statistics from Training Set Naive Bayes (Accuracy):
       Mean       Max  Range       Mean          SDev
0   0.964781  0.964781     0.0   0.964781   3.330669e-16
Training Data Confusion Matrix (Naive Bayes):
[[454    4]
 [ 29  450]]
Test Data Confusion Matrix:
[[189   27]
 [ 49  197]]
Test Data Confusion Matrix (Random Forest):
[[162   54]
 [ 19  227]]
```

When evaluating the results, the neural network has a tendency to overfit to training data, but its test performance is very similar to logistic regression, although slightly less accurate. When assessing the results of the naïve Bayes classifier and the random forest classifier, we observe roughly similar AUC scores, with only a difference in false positives and true positives as the trade-off that we must accept. In this instance, it is important to consider our objective.

If we are using these algorithms to label the reviews that users input, and then perform analytics on top of these reviews, we want to maximize the accuracy rate, or seek models with the highest true positive rate and true negative rate. In the instance of spam detection, we likely want the model that has the best ability to properly classify spam from normal mail.

I have introduced and applied the bag-of-words schemes in both the logistic model and the naïve Bayes classifier. This brings us to the final part of this section, in which I discuss their relative advantages and disadvantages. You should be aware of this so as to not waste time altering subpar solutions. The major advantage of BoW is that it is a relatively straightforward algorithm that allows you to quickly turn text into a format interpretable by a machine learning algorithm, and to attack NLP problems directly.

The largest disadvantage of BoW is its relative simplicity. BoW does not account for the context of words, and as such, it does not make it the ideal feature extraction method for more complex NLP tasks. For example, "4" and "four" are considered semantically indistinguishable, but in BoW, they are considered two different words altogether. When we expand this to phrases, "I went to college for four years," and "For 4 years, I attended a university," are treated as orthogonal vectors. Another example of a BoW shortcoming is that it cannot distinguish the ordering of words. As such, "I am stupid" and "Am I stupid" appear as the same vector.

Because of these shortcomings, it is appropriate for us to utilize more advanced models, such as word embeddings, for these difficult problems, which are discussed in detail in the next chapter.

Summary

This brings us to the end of Chapter 3! This chapter tackled working with text data in document classification problems. You also became familiar with two BoW feature extraction methods.

Let's take a moment to go over some of the most important lessons from this chapter. Just as with traditional machine learning, you must define the type of problem and analyze data. Is this simply document classification? Are we trying to find synonyms? We have to answer these questions before tackling any other steps.

The removal of stop words, grammatical tokens, and frequent words improves the accuracy of our algorithms. Not every word in a document is informative, so you should know how to remove the noise. That said, over-selecting features can be detrimental to our model's success, so you should be aware of this too!

Whenever you are working with a machine learning problem, within or outside the NLP domain, you must *establish a baseline solution and then improve if necessary!* I suggest that you always start a deep learning problem by seeing how the solution appears, such as with a logistic regression. Although it is my goal to teach you how to apply deep learning to NLP-based problems, there is no reason to use overly complex methods where less complex methods will do better or equally as well (unless you like to practice your deep learning skills).

Finally, while preprocessing methods are useful, BoW-based models are best utilized with document classification. For more advanced NLP problems, such as sentiment analysis, understanding semantics, and similarly abstract problems, BoW likely will not yield the best results.

CHAPTER 4

Topic Modeling and Word Embeddings

Now that you have had an introduction to working with text data, let's dive into one of the more advanced feature extraction algorithms. To accomplish some of the more difficult problems, it is reasonable for me to introduce you to other techniques to approach NLP problems. We will move through Word2Vec, Doc2Vec, and GloVe.

Topic Model and Latent Dirichlet Allocation (LDA)

Topic models are a method of extracting information from bodies of text to see what "topics" occur across all the documents. The intuition is that we expect certain topics to appear more in relevant documents and not as much in irrelevant documents. This might be useful when using the topics we associate with a document as keywords for better and more intuitive search, or when using it for shorthand summarization. Before we apply this model, let's talk about how we actually extract topics.

© Taweh Beysolow II 2018
T. Beysolow II, *Applied Natural Language Processing with Python*,
https://doi.org/10.1007/978-1-4842-3733-5_4

Latent Dirichlet allocation (LDA) is a generative model developed in 2003 by David Blei, Andrew Ng, and Michael I. Jordan. In their paper, they highlight the shortcomings of TFIDF. Most notably, TFIDF is unable to understand the semantics of words, or the position of a word in text. This led to the rise of LDA. LDA is a generative model, meaning that it outputs all the possible outcomes for a given phenomenon. Mathematically, we can describe the assumptions as follows:

1. Choose $N \sim Poisson(\xi)$ (a sequence of N words within a document have a Poisson distribution)

2. Choose $\theta \sim Dir(\alpha)$ (a parameter θ has a Dirichlet distribution)

3. For each of the N words (w_n):

   • Choose topic $z_n \sim Multinomial(\theta)$ (Each topic z_n has a multinomial distribution.)

   • Choose w_n from $p\,(w_n\,|\,z_n,\,\beta)$, a multinomial probability conditional on topic z_n. (Each topic is represented as a distribution over words, where a probability is generated from the probability of the nth word, conditional upon the topic as well as β where $\beta_{ij} = p(w^j = 1|z^i = 1)$ with dimensions $k \times V$.)

     β = probability of a given word, V = number of words in the vocabulary, k = the dimensionality of the Dirichlet distribution, θ = the random variable sampled from the probability simplex.

Let's discuss some of the distributions utilized in these assumptions. The Poisson distribution represents events that occur in a fixed time or space and at a constant rate, independently of the time since the last event. An example of this distribution is a model of the number of people who call a pizzeria for delivery during a given period of time. The multinomial distribution is the k-outcome generalization of the binomial distribution; in other words, the same concept as the binomial distribution but expanded to cases where there are more than two outcomes.

Finally, the Dirichlet distribution is a generalization of the beta distribution, but expanded to handle multivariate data. The beta distribution is a distribution of probabilities.

LDA assumes that (1) words are generated from topics, which have fixed conditional distributions, and (2) that the topics within a document are infinitely exchangeable, which means that the joint probability distribution of these topics is *not affected* by the order in which they are represented. Reviewing statements 1 and 2 allows us to state that words within a topic *are not* infinitely exchangeable.

Let's discuss parameter θ (drawn from the Dirichlet distribution), which the dimensionality of the distribution, k, is utilized. We assume that k is known and fixed, and that k-dimensional Dirichlet random variable θ can take any values in the $(k-1)$ probability simplex. Here, we define the probability simplex as the area of the distribution that we draw the random variable from, graphically represented as a multidimensional triangle with $k + 1$ vertices. The probability distribution on the simplex itself can be represented as follows:

$$p(\theta|\alpha) = \left(\frac{\Gamma\left(\sum_{i=1}^{k}\alpha_i\right)}{\prod_{i=1}^{k}\Gamma(\alpha_i)} \right) \theta_1^{\alpha_1-1},\dots,\theta_k^{\alpha_1-1} \tag{4.1}$$

α = k-vector of positive real valued numbers. $\Gamma(x)$ = gamma function.

Subsequently, we define the joint distribution of a mixture of topics, as follows:

$$p(\theta, \mathbf{z}, \mathbf{w}|\alpha, \beta) = p(\theta|\alpha) \prod_{n=1}^{N} p(z_n|\theta)p(w_n|z_n, \beta) \tag{4.2}$$

Therefore, a given sequence of words and topics must have the following form:

$$p(\mathbf{w}, \mathbf{z}) = \int p(\theta) \left(\prod_{n=1}^{N} p(z_n | \theta) \right) p(w_n | z_n) \tag{4.3}$$

In the LDA paper, the authors provide a useful illustration for Equations 4.1, 4.2, and 4.3, as shown in Figure 4-1.

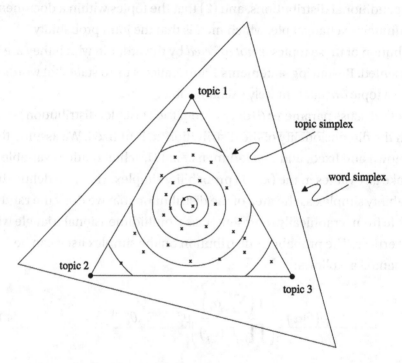

Figure 4-1. *Topic and word simplexes*

The example given in the LDA paper describes Figure 4-1 as an illustration of a topic simplex inside a word simplex comprised of three words. Each of the simplex points represent a given word and topic, respectively.

Before we complete our discussion on the theory behind LDA, let's re-create all the work in Python. Thankfully, scikit-learn provides an implementation of LDA that we will utilize in the upcoming example.

Topic Modeling with LDA on Movie Review Data

Next, we look at the same movie review data that we used in our document classification example. The following is an example of some of the code that we will utilize to first create out topic model. We'll start with an implementation in sklearn.

```python
def create_topic_model(model, n_topics=10, max_iter=5, min_
df=10, max_df=300, stop_words='english', token_pattern=r'\w+'):
    print(model + ' topic model: \n')
    data = load_data()[0]
    if model == 'tf':
        feature_extractor = CountVectorizer(min_df=min_df, max_
            df=max_df, stop_words=stop_words, token_pattern=r'\w+')
    else:
        feature_extractor = TfidfVectorizer(min_df=min_df, max_
            df=max_df, stop_words=stop_words, token_pattern=r'\w+')
    processed_data = feature_extractor.fit_transform(data)
```

We load the movie reviews that we used in Chapter 3 for the classification problem. In this example, we will imagine that we want to make a topic model for a given number of movie reviews.

Note   We are importing the `load_data()` function from a previous file. To execute the `lda_demo.py` file, use a relative import from the `code_applied_nlp_python` directory, and execute the following command: `'python -m chapter4.topic_modeling'`

We load the data in with our function and then prepare it for input to the LDA fit_transform() method. Like other NLP problems, we cannot put raw text into any of our algorithms; we must always preprocess it in some form. However, for producing a topic model, we will utilize both the term frequency and the TFIDF algorithms, but mainly to compare results.

Let's move through the rest of the function.

```
lda_model = LatentDirichletAllocation(n_topics=n_topics,
learning_method='online', learning_offset=50., max_iter=max_
iter, verbose=1)
lda_model.fit(processed_data)
tf_features = feature_extractor.get_feature_names()
print_topics(model=lda_model, feature_names=tf_features, n_top_
words=n_top_words)
```

When we execute the following function, we get this as our output:

```
tf topic model:
Topic #0: libby fugitive douglas sarah jones lee detective
double innocent talk
Topic #1: beatty joe hanks ryan crystal niro fox mail
kathleen shop
Topic #2: wars phantom lucas effects menace neeson jedi anakin
special computer
Topic #3: willis mercury simon rising jackal bruce ray lynch
baseball hughes
Topic #4: godzilla broderick redman bvoice kim michael
bloomington mission space york
Topic #5: planet apes joe sci fi space ape alien gorilla newman
Topic #6: d american fun guy family woman day 11 james bit
Topic #7: bond brosnan bottle message blake theresa pierce
tomorrow dies crown
```

```
Topic #8: van spielberg amistad gibson en home american kevin
ending sense
Topic #9: scream 2 wild williamson horror smith kevin arquette
sidney finn
```

Being that this is movie data, we can see that the topics refer to both the movie and the context surrounding it. For example, topic #4 lists "Godzilla" (ostensibly a character) and "Broderick" (ostensibly an actor). We can also produce topic models utilizing other feature extraction methods.

Now let's look at the results of the topic model when we use the TFIDF feature extractor.

```
tfidf topic model:
Topic #0: libby driver jacket attending terrorists tends finn
doom tough parodies
Topic #1: godzilla beatty douglas arthur christ technology
burns jesus york cases
Topic #2: wars lucas episode menace jar niro jedi darth anakin
phantom
Topic #3: sub theron genre keaton cooper victor rita irene
dating rules
Topic #4: midnight kim stiller mulan spice newman disney junkie
troopers strange
Topic #5: clooney palma kevin pacino snake home toy woody
pfeiffer space
Topic #6: anna disney jude carpenter men wrong siege lee king
family
Topic #7: scream got mail bond hanks book performances summer
cute dewey
Topic #8: en van z n er met reese die fallen lou
Topic #9: family american effects home guy woman michael
original 10 james
```

There are similar results, although we get slightly different results for some of the topics. In some ways, the TFIDF model can be less interpretable than the term-frequency model.

Before we move forward, let's discuss how to utilize the LDA model with a new package. Gensim is a machine learning library that is heavily focused on applying machine learning and deep learning to NLP tasks. The following is code that utilizes this package in the gensim_topic_ model() function:

```
def gensim_topic_model():

    def remove_stop_words(text): (1)
        word_tokens = word_tokenize(text.lower())
        word_tokens = [word for word in word_tokens if word not
        in stop_words and re.match('[a-zA-Z\-][a-zA-Z\-]{2,}',
        word)]
        return word_tokens

    data = load_data()[0]
    cleaned_data = [remove_stop_words(data[i]) for i in
    range(0, len(data))]
```

When using this package, the Gensim LDA implementation expects a different input than the gensim implementation, although it still requires preprocessing. When looking at function, we have to remove stop words using a proprietary function, as we did earlier in Chapter 3. In addition to this, we should be mindful to remove words that appear too frequently, and not frequently enough. Thankfully, Gensim provides a method within the corpora.Dictionary() function to do this, as shown here:

```
dictionary = gensim.corpora.Dictionary(cleaned_data)
dictionary.filter_extremes(no_below=100, no_above=300)
corpus = [dictionary.doc2bow(text) for text in cleaned_data]
lda_model = models.LdaModel(corpus=corpus, num_topics=n_topics,
id2word=dictionary, verbose=1)
```

Similar to the scikit-learn method, we can filter objects based on document frequency. The preprocessing steps we are taking here are slightly different than those present in the `sklearn_topic_model()` function, which will become central to our discussion at the end of this section. Similar to what you saw before, what seems like a minor change in preprocessing steps can lead to a drastically different outcome.

We execute the `gensim_topic_model()` function and get the following result:

```
Gensim LDA implemenation:
Topic #0: 0.116*"movie" + 0.057*"people" + 0.051*"like" +
0.049*"good" + 0.041*"well" + 0.038*"film" + 0.037*"one" +
0.037*"story" + 0.033*"great" + 0.028*"new"
Topic #1: 0.106*"one" + 0.063*"movie" + 0.044*"like" +
0.043*"see" + 0.041*"much" + 0.038*"story" + 0.033*"little" +
0.032*"good" + 0.032*"way" + 0.032*"get"
Topic #2: 0.154*"film" + 0.060*"one" + 0.047*"like" +
0.039*"movie" + 0.037*"time" + 0.032*"characters" +
0.031*"scene" + 0.028*"good" + 0.028*"make" + 0.027*"little"
Topic #3: 0.096*"film" + 0.076*"one" + 0.060*"even" +
0.053*"like" + 0.051*"movie" + 0.040*"good" + 0.036*"time" +
0.033*"get" + 0.030*"would" + 0.028*"way"
Topic #4: 0.079*"film" + 0.068*"plot" + 0.058*"one" +
0.057*"would" + 0.049*"like" + 0.039*"two" + 0.038*"movie" +
0.036*"story" + 0.035*"scenes" + 0.033*"much"
Topic #5: 0.136*"film" + 0.067*"movie" + 0.064*"one" +
0.039*"first" + 0.037*"even" + 0.037*"would" + 0.036*"time" +
0.035*"also" + 0.029*"good" + 0.027*"like"
Topic #6: 0.082*"movie" + 0.072*"get" + 0.068*"film" +
0.059*"one" + 0.046*"like" + 0.036*"even" + 0.035*"know" +
0.027*"much" + 0.027*"way" + 0.026*"story"
```

```
Topic #7: 0.131*"movie" + 0.097*"film" + 0.061*"like" +
0.045*"one" + 0.032*"good" + 0.029*"films" + 0.027*"see" +
0.027*"bad" + 0.025*"would" + 0.025*"even"
Topic #8: 0.139*"film" + 0.060*"movie" + 0.052*"like" +
0.044*"story" + 0.043*"life" + 0.043*"could" + 0.041*"much" +
0.032*"well" + 0.031*"also" + 0.030*"time"
Topic #9: 0.116*"film" + 0.091*"one" + 0.059*"movie" +
0.035*"two" + 0.029*"character" + 0.029*"great" + 0.027*"like"
+ 0.026*"also" + 0.026*"story" + 0.026*"life"
```

So far, the results from the scikit-learn implementation of LDA using term frequency as our feature extractor has given the most interpretable results. Most of the results are homogenous, which might not lead to much differentiation, making the results from this less useful.

Using this same data set, let's utilize another topic extraction model.

Non-Negative Matrix Factorization (NMF)

Non-negative matrix factorization (NMF) is an algorithm that takes a matrix and returns two matrices that have no non-negative elements. NMF is closely related to matrix factorization, except NMF only receives non-negative values (0 and anything above 0).

We want to utilize NMF rather than another type of matrix factorization because we need positive coefficients, as is the case when using LDA. We can describe the process with the following mathematical formula:

$$V = WH$$

The matrix, V, is the original matrix that we input to the data. The two matrices that we output are W and H. In this example, let's assume matrix V has 1000 rows and 200 columns. Each row represents a word and each column represents a document. Therefore, we have a 1000-word vocabulary featured across 200 documents. As it relates to the preceding equation, V is an $m{\times}n$ matrix, W is an $m{\times}p$ matrix, and H is a $p{\times}n$ matrix. W is a features matrix.

Let's say that we would like to find five features such that we generate matrix W with 1000 rows and 5 columns. Matrix H subsequently has a shape equivalent to 5 rows and 200 columns. When we perform matrix multiplication on W and H, we yield matrix V with 1000 rows and 200 columns, equivalent to the dimensionality described earlier. We consider that each document is built from a number of hidden features, which NMF would therefore generate. The following is the scikit-learn implementation of NMF that we will utilize for this example:

```
def nmf_topic_model():

    def create_topic_model(model, n_topics=10, max_iter=5,
    min_df=10,
                            max_df=300, stop_words='english',
                            token_pattern=r'\w+'):
        print(model + ' NMF topic model: ')
        data = load_data()[0]
        if model == 'tf':
            feature_extractor = CountVectorizer(min_df=min_df,
            max_df=max_df,
                                    stop_words=stop_words,
                                    token_pattern=token_pattern)
        else:
            feature_extractor = TfidfVectorizer(min_df=min_df,
            max_df=max_df,
                                    stop_words=stop_words,
                                    token_pattern=token_pattern)

        processed_data = feature_extractor.fit_transform(data)
        nmf_model = NMF(n_components=n_components,
        max_iter=max_iter)
        nmf_model.fit(processed_data)
        tf_features = feature_extractor.get_feature_names()
```

```
    print_topics(model=nmf_model, feature_names=tf_
    features, n_top_words=n_topics)

  create_topic_model(model='tf')
```

We invoke the NMF topic extraction in virtually the same manner that we invoke the LDA topic extraction model. Let's look at the output of both the term frequency preprocessed data and the TFIDF preprocessed data.

```
tf NMF topic model:
Topic #0: family guy original michael sex wife woman r men play
Topic #1: jackie tarantino brown ordell robert grier fiction
pulp jackson michael
Topic #2: jackie hong drunken master fu kung chan arts martial ii
Topic #3: scream 2 williamson horror sequel mr killer sidney
kevin slasher
Topic #4: webb jack girl gives woman ll male killed sir talking
Topic #5: musical musicals jesus death parker singing woman
nation rise alan
Topic #6: bulworth beatty jack political stanton black warren
primary instead american
Topic #7: godzilla effects special emmerich star york computer
monster city nick
Topic #8: rock kiss city detroit got music tickets band
soundtrack trying
Topic #9: frank chicken run shannon ca mun html sullivan
particularly history
```

The following is the TFIDF NMF topic model:

```
Topic #0: 10 woman sense james sex wife guy school day ending
Topic #1: scream horror williamson 2 sidney craven stab killer
arquette 3
Topic #2: wars phantom jedi lucas menace anakin jar effects
darth gon
```

Topic #3: space deep alien ship armageddon harry effects godzilla impact aliens
Topic #4: disney mulan animated joe voice toy animation apes mermaid gorilla
Topic #5: van amistad spielberg beatty cinque political slavery en slave hopkins
Topic #6: carpenter ott ejohnsonott nuvo subscribe reviews johnson net mail e
Topic #7: hanks joe ryan kathleen mail shop online fox tom meg
Topic #8: simon sandler mercury adam rising willis wedding vincent kevin julian
Topic #9: murphy lawrence martin eddie ricky kit robbins miles claude police

Before we evaluate the results and have a more thorough discussion on both methods, let's focus on visualizing the results. In the preceding example, we've reasonably reduced the complexity so that users can assess the different topics within the analyzed documents. However, this isn't as helpful when we want to look at larger amounts of data and make inferences from this topic model relatively quickly.

We'll begin with what I believe is a useful plot, supplied by pyLDAvis. This software is extremely useful and works relatively easily when used with a Jupyter notebook, which are excellent for code visualization and results presentation. It is common to utilize a Jupyter notebook when using a virtual machine instance from either Amazon Web Services (AWS) or Google Cloud.

Note   For those of you who have not worked with Google Cloud or AWS, I recommend these tutorials: Google Compute Engine: www.youtube.com/watch?v=zzMCKv1g5z0 AWS: www.youtube.com/watch?v=q1vVedHbkAY

Set up an instance and start a Jupyter notebook. We will make some minor adjustments for running this on your local machine to running it in the cloud. In this example, the scikit-learn implementations—given the preprocessing algorithms provided—make gleaning interpretable topics much easier than the Gensim model. Although it gives more flexibility and has a lot of features, Gensim requires you to fine-tune the preprocessing steps from scratch. If you have the time to build results from scratch, this is not a problem; however, keep this in mind when building your own application, and consider the difficulties of having to use this method in Gensim.

In this demo, NMF and LDA typically give similar results; however, the choice of one model vs. the other is often relative to the way we conceive of the data. LDA assumes that topics are infinitely exchangeable, but the words within a topic are not. As such, if we are not concerned about the topic probability per document remaining fixed (it assumedly would not be, as not all documents contain the same topics across large corpuses), LDA is a better choice. NMF might be a better choice if we have a heavy degree of certainty with respect to fixed topic probability and the data set is considerably smaller. Again, these statements should be taken in consideration when evaluating the results of the respective topic models, as with all machine learning problems.

Let's discuss a more advanced modeling technique that plays a role in sentiment analysis (in addition to more advanced NLP tasks): word embeddings. We begin by discussing a body of algorithms: Word2Vec.

Word2Vec

Tomas Mikolov, Ilya Sutskever, Kai Chen, Greg Corrado, and Jeffrey Dean are credited with creating Word2Vec in 2014 while working at Google. Word2Vec represents a significant step forward in NLP-related tasks, as it provides a method for finding vector representations of words and phrases, and it can be expanded as much as the documents.

First, let's examine the Skip-Gram model, which is a shallow neural network whose objective is to predict a word in a sequence based on the words around it. Let's take the following sequence of training words:

$$w = \left(w_1, w_2, \ldots, w_T\right)$$

The objective function is the average log probability, represented by the following:

$$\frac{1}{T}\sum_{t=1}^{T}\sum_{-c \leq j \leq c, j \neq 0} \log p(w_{t+j}|w_t)$$

$$p(w_{t+j}|w_t) = p(w_o|w_I) = \frac{\exp\left(v'_{w_O}{}^T v_{wi}\right)}{\sum_{w=1}^{W}\exp\left(v'_{w}{}^T v_{wi}\right)}$$

c = size of the training context, T = the total number of training words, t = index position of the current word, j = the window that determines which word in the sequence we are looking, w_t = center word of the sequence, and W = number of words in the vocabulary.

Before we move on, it's important that you understand the formula and how it explains what the model does. An n-gram is a continuous grouping of n words. A Skip-Gram is a generalization of an n-gram, such that we have groupings of words, but they no longer need to be continuous; that is, we can skip words to create Skip-Grams. Mathematically, we typically define k-skip-n-grams as follows:

$$\left\{w_{i_1}, w_{i_2}, \ldots, w_{i_n} \Big| \sum_{j=1}^{n} i_j - i_{j-1} < k\right\}$$

Let's assume the following is an input to the k-skip-n-gram model: "The cat down the street"

Let's also assume that we are seeking to create a 2-skip-bi-gram model. As such, the training examples are as follows:

- "The, cat", "The, down", "The, the", "The, street"
- "cat, the", "cat, down", "cat, the", "cat, street"

- "down, the", "down, cat", "down, the", "down, street"

- "the, The", "the, cat", "the, down", "the, street"

- "street, The", "street, cat", "street, down", "street, the"

So now you understand how the input data is represented as words.

Let's discuss how we represent these words with respect to a neural network. The input layer for the Skip-Gram model is a one-hot encoded vector with W components. In other words, every element of the vector is representative of a word in the vocabulary. The Skip-Gram architecture is graphically represented in Figure 4-2.

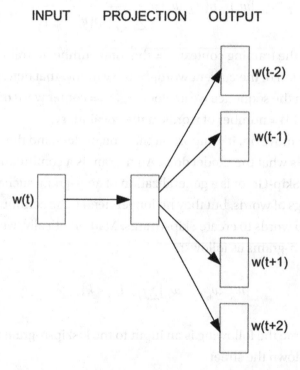

Figure 4-2. Skip-Gram model architecture

The goal of the neural network is to predict the word that has the highest probability of coming next in the input sequence. This is precisely why we want to use softmax, and ultimately how you intuitively understand the formula. *We want to predict the word that is most probable given the input of the words, and we are calculating this probability based on the entirety of the input and output sequences observed by the neural network.*

Be that as it may, we have a minor problem. Softmax computation scales proportionally to the input size, which bodes poorly for this problem because accurate results will likely require large vocabulary sizes for training data. Hence, it is often suggested that we use an alternative method. One of the methods often referenced is negative sampling. Negative sampling is defined in the following equation:

$$\log \sigma \left(v_{wo}'^{T} v_{wi} \right) + \sum_{i=1}^{k} \mathbb{E}_{wi} \sim P_n(w) \left[\log \sigma \left(-v_{wi}^{T} v_{wi} \right) \right]$$

Negative sampling achieves a cheaper computation than the softmax activation function by approximating its output. More precisely, we are only going to change K number of weights in the word embedding rather than computing them all. The Word2Vec paper suggests to sample with 5 to 20 words in smaller data sets, but 2 to 5 words in larger data sets can achieve positive results.

Beyond the training of the word embedding, what are we actually going to use it for? Unlike many neural networks, the main objective is not necessarily to use it for the purpose of prediction, but rather to obtain the trained hidden layer weight matrix. *The hidden layer weight matrix is our trained word embedding.* Once this hidden layer is trained, certain words cluster in areas of vector space, where they share similar contexts.

Example Problem 4.2: Training a Word Embedding (Skip-Gram)

Let's display the power of Word2Vec by working through a demo example, in Gensim and in TensorFlow. The following is some of the code that begins our implementation of the TensorFlow Word2Vec Skip-Gram model:

```python
def remove_non_ascii(text):
    return "".join([word for word in text if ord(word) < 128])

def load_data(max_pages=100):
    return_string = StringIO()
    device = TextConverter(PDFResourceManager(), return_string,
    codec='utf-8', laparams=LAParams())
    interpreter = PDFPageInterpreter(PDFResourceManager(),
    device=device)
    filepath = file('/Users/tawehbeysolow/Desktop/applied_nlp_
    python/datasets/economics_textbook.pdf', 'rb')
    for page in PDFPage.get_pages(filepath, set(),
    maxpages=max_pages, caching=True, check_extractable=True):
        interpreter.process_page(page)
    text_data = return_string.getvalue()
    filepath.close(), device.close(), return_string.close()
    return remove_non_ascii(text_data)
```

For our example problem, we will utilize the PDFMiner Python module. For those of you who often parse data in different forms, this package is highly recommended. PDF data is notorious in parsing, as it is often filled with images and metadata that makes preprocessing the data a hassle. Thankfully, PDFMiner takes care of most of the heavy lifting, making our primary concerns only cleaning out stop words, grammatical characters, and other preprocessing steps, which are relatively straightforward. For this problem, we will read data from an economics textbook.

```
def gensim_preprocess_data():
    data = load_data()
    sentences = sent_tokenize(data)
    tokenized_sentences = list([word_tokenize(sentence) for
    sentence in sentences])
    for i in range(0, len(tokenized_sentences)):
        tokenized_sentences[i] = [word for word in tokenized_
        sentences[i] if word not in punctuation]
    return tokenized_sentences
```

We now move to tokenizing the data based on sentences. *Do not remove punctuation before this step.* The NLTK sentence tokenizer relies on punctuation to determine where to split data based on sentences. If this is removed, it can cause you to debug something rather trivial. Regardless, the next format the data should take is that of a list, where every entry is a sentence whose words are tokenized, such that the words appear as follows:

```
[['This', 'text', 'adapted', 'The', 'Saylor', 'Foundation',
'Creative', 'Commons', 'Attribution-NonCommercial-ShareAlike',
'3.0', 'License', 'without', 'attribution', 'requested',
'works', 'original', 'creator', 'licensee'], ['Saylor',
'URL', 'http', '//www.saylor.org/books', 'Saylor.org', '1',
'Preface', 'We', 'written', 'fundamentally', 'different',
'text', 'principles', 'economics', 'based', 'two', 'premises',
'1'], ['Students', 'motivated', 'study', 'economics', 'see',
'relates', 'lives'], ['2'], ['Students', 'learn', 'best',
'inductive', 'approach', 'first', 'confronted', 'question',
'led', 'process', 'answer', 'question'], ['The', 'intended',
'audience', 'textbook', 'first-year', 'undergraduates',
'taking', 'courses', 'principles', 'macroeconomics',
'microeconomics'], ['Many', 'may', 'never', 'take',
'another', 'economics', 'course'], ['We', 'aim', 'increase',
'economic', 'literacy', 'developing', 'aptitude', 'economic',
```

'thinking', 'presenting', 'key', 'insights', 'economics',
'every', 'educated', 'individual', 'know'], ['Applications',
'ahead', 'Theory', 'We', 'present', 'theory', 'standard',
'books', 'principles', 'economics'], ['But', 'beginning',
'applications', 'also', 'show', 'students', 'theory',
'needed'], ['We', 'take', 'kind', 'material', 'authors', 'put',
'applications', 'boxes', 'place', 'heart', 'book'], ['Each',
'chapter', 'built', 'around', 'particular', 'business',
'policy', 'application', 'microeconomics', 'minimum', 'wages',
'stock', 'exchanges', 'auctions', 'macroeconomics', 'social',
'security', 'globalization', 'wealth', 'poverty', 'nations']

Now that we have finished the preprocessing of our data, we can work with the Gensim implementation of the Skip-Gram model.

```
def gensim_skip_gram():
    sentences = gensim_preprocess_data()
    skip_gram = Word2Vec(sentences=sentences, window=1,
    min_count=10, sg=1)
    word_embedding = skip_gram[skip_gram.wv.vocab] (1)
```

Invoking the Skip-Gram model is relatively straightforward, and the training of the model is taken care of for us as well. The training process of a Skip-Gram model mimics that of all neural networks, in that we pass an input through all the layers and then backpropagate the error through each of the respective weights in each layer, updating them until we have reached a loss tolerance threshold or until the maximum number of epochs has been reached. Once the word embedding has been trained, we obtain the weight matrix by indexing the model with the wv.vocab attribute of the model itself.

Now, let's discuss visualizing the words as vectors.

```
pca = PCA(n_components=2)
word_embedding = pca.fit_transform(word_embedding)
```

```
#Plotting results from trained word embedding
plt.scatter(word_embedding[:, 0], word_embedding[:, 1])
word_list = list(skip_gram.wv.vocab)
for i, word in enumerate(word_list):
    plt.annotate(word, xy=(word_embedding[i, 0],
    word_embedding[i, 1]))
```

Word embeddings are output in dimensions that are difficult to visualize in their raw formats. As such, we need to find a way to reduce the dimensionality of this matrix, while also retaining all the variance and attributes of the original data set. A preprocessing method that does this is principal components analysis (PCA). Briefly, PCA transforms a matrix so that it returns an eigen-decomposition called eigenvectors, in addition to eigenvalues. For the sake of showing a two-dimensional plot, we want to create a transformation that yields *two* principal components. It is important to remember that these principal components are *not* exactly the same as the original matrix, but an orthogonal transformation of the word embedding that is related to it. Figure 4-3 illustrates the matrix.

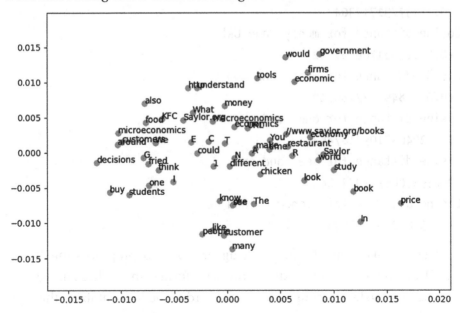

Figure 4-3. *Skip-Gram word embeddings generated via Gensim*

In the vector space, words that are closer to one another appear in similar contexts, and words that are further away from each other are more dissimilar in respect to the contexts in which they appear. Cosine similarity is a common method for measuring this. Mathematically, cosine distance is described as follows:

$$\cos(\theta) = \frac{\mathbf{A} * \mathbf{B}}{\|\mathbf{A}\|\|\mathbf{B}\|}$$

We intuitively describe cosine similarity as the sum of the product of all the respective elements of two given vectors, divided by the product of their Euclidean norms. Two vectors that have a 0-degree difference yield a cosine similarity of 1; whereas two vectors with a 90-degree difference yield a cosine similarity of 0. The following is an example of some of the cosine distances between different word vectors:

```
Cosine distance for people  and Saylor
 -0.10727774727479297
Cosine distance for URL  and people
 -0.137377917173043
Cosine distance for money  and URL
 -0.03124461706797222
Cosine distance for What  and money
 -0.007384979727807199
Cosine distance for one  and What
 0.022940581092187863
Cosine distance for see  and one
 0.05983065381073224
Cosine distance for economic  and see
 -0.0530102968258333
```

Gensim takes care of some of the uglier aspects of preprocessing the data. However, it is useful to know how to perform some of these things from scratch, so let's try implementing a word embedding utilizing the same data, except this time we will do it in TensorFlow.

Let's walk through a toy implementation to ensure that you are aware of what the model is doing, and then walk through an implementation that is easier to deploy.

```
def tf_preprocess_data(window_size=window_size):

    def one_hot_encoder(index, vocab_size):
        vector = np.zeros(vocab_size)
        vector[index] = 1
        return vector

    text_data = load_data()
    vocab_size = len(word_tokenize(text_data))
    word_dictionary = {}
    for index, word in enumerate(word_tokenize(text_data)):
        word_dictionary[word] = index

    sentences = sent_tokenize(text_data)
    tokenized_sentences = list([word_tokenize(sentence) for
sentence in sentences])
    n_gram_data = []
```

We must prepare the data slightly differently for TensorFlow than we did for Gensim. The Gensim Word2Vec method takes care of most of the back-end things for us, but it is worthwhile to implement a simple proof of concept from scratch and walk through the algorithm.

We begin by making a dictionary that matches a word with an index number. This index number forms the position in our one-hot encoded input and output vectors.

Let's continue preprocessing the data.

```
#Creating word pairs for word2vec model
    for sentence in tokenized_sentences:
        for index, word in enumerate(sentence):
            if word not in punctuation:
```

```
            for _word in sentence[max(index - window_size, 0):
                                   min(index + window_size,
                                   len(sentence)) + 1]:
                if _word != word:
                    n_gram_data.append([word, _word])
```

The preceding section of code effectively creates our n-grams, and ultimately simulates how the Skip-Gram model convolves over a sentence in such a way that it can predict the following word with the highest probability. We then create an $m \times n$ matrix, where m is the number of words in our input sequences, and n is the number words in the vocabulary.

```
#One-hot encoding data and creating dataset intrepretable by
skip-gram model
x, y = np.zeros([len(n_gram_data), vocab_size]),
np.zeros([len(n_gram_data), vocab_size])

for i in range(0, len(n_gram_data)):
    x[i, :] = one_hot_encoder(word_dictionary[n_gram_data[i]
    [0]], vocab_size=vocab_size)
    y[i, :] = one_hot_encoder(word_dictionary[n_gram_data[i]
    [1]], vocab_size=vocab_size)

return x, y, vocab_size, word_dictionary
```

Moving forward to the function that we will use to construct our Skip-Gram model, we begin by loading the data and the vocabulary size and word dictionary. As with other neural network models, we instantiate the placeholders, variables, and weights. Per the Skip-Gram model diagram shown in Figure 4-2, we only need to contain a hidden and an output weight matrix.

```python
def tensorflow_word_embedding(learning_rate=learning_rate,
embedding_dim=embedding_dim):
    x, y, vocab_size, word_dictionary = tf_preprocess_data()

    #Defining tensorflow variables and placeholder
    X = tf.placeholder(tf.float32, shape=(None, vocab_size))
    Y = tf.placeholder(tf.float32, shape=(None, vocab_size))

    weights = {'hidden': tf.Variable(tf.random_normal([vocab_
    size, embedding_dim])),
                'output': tf.Variable(tf.random_
                normal([embedding_dim, vocab_size]))}

    biases = {'hidden': tf.Variable(tf.random_
    normal([embedding_dim])),
                'output': tf.Variable(tf.random_normal([vocab_
                size]))}

    input_layer = tf.add(tf.matmul(X, weights['hidden']),
    biases['hidden'])
    output_layer = tf.add(tf.matmul(input_layer,
    weights['output']), biases['output'])
```

In Chapter 5, we walk through implementing negative sampling. However, because the number of examples that we are using here is relatively miniscule, we can get away with utilizing the regular implementation of softmax as provided by TensorFlow. Finally, we execute our graph, as with other TensorFlow models, and observe the results shown in Figure 4-4.

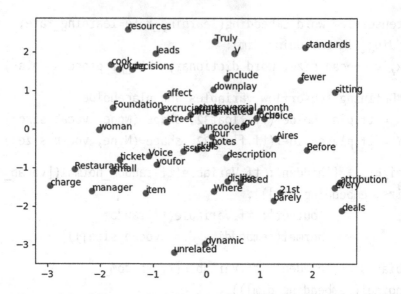

Figure 4-4. *Word vectors from toy implementation of Skip-Gram*

```
Cosine distance for dynamic  and limited
 0.4128825113896724
Cosine distance for four  and dynamic
 0.2833843609582811
Cosine distance for controversial  and four
 0.3266445485300576
Cosine distance for hanging  and controversial
 0.37105348488163503
Cosine distance for worked  and hanging
 0.44684699747383416
Cosine distance for Foundation  and worked
 0.3751656692569623
```

Again, the implementations provided here are *not* final examples of what well-trained word embeddings necessarily looks like. We will tackle that task more specifically in Chapter 5, as data collection is largely the issue that we must discuss in greater detail. However, the Skip-Gram model is only one of the word embeddings that we will likely encounter.

We now will continue our discussion by tackling the continuous bag-of-words model.

Continuous Bag-of-Words (CBoW)

Similar to a Skip-Gram model, a continuous bag-of-words model (CBoW) is training on the objective of predicting a word. Unlike the Skip-Gram model, however, we are not trying to predict the next word in a given sequence. Instead, we are trying to predict some center word based on the context around the target label. Let's imagine the following input data sentence:

"The boy walked to the red house"

In the context of the CBoW model, we could imagine that we would have an input vector that appeared as follows:

"The, boy, to, the, red, house"

Here, "walked" is the target that we are trying to predict. Visually, the CBoW model looks like Figure 4-5.

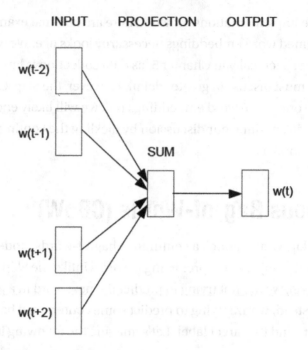

Figure 4-5. *CBoW model representation*

Each word in the input is represented in a single one-hot encoded vector.
Similar to the Skip-Gram model, the length of the input vector is equal to
the number of words in the vocabulary. When evaluating our input data, a
value of "1" is for the words that are present and a "0" is for the words that
are not present. In Figure 4-5, we are predicting a target word, w_t, based
on the words w_t-2, w_ t-1, w_ t+1, and w_t+2.

We then perform a weighted sum operation on this input vector with
weight and bias matrices that pass these values to the projection layer,
which is similar to the projection layer featured in the Skip-Gram model.
Finally, we predict the class label with another weighted sum operation
with the output weight and bias matrices in addition to utilizing a softmax
classifier. The training method is the same as the one used in the Skip-
Gram model.

Next, let's work with a short example utilizing Gensim.

Example Problem 4.2: Training a Word Embedding (CBoW)

The Gensim implementation of CBoW requires that only a single parameter is changed, as shown here:

```
cbow = Word2Vec(sentences=sentences, window=skip_gram_window_
size, min_count=10, sg=0)
```

We invoke this method and observe the results in the same manner that we did for the Skip-Gram model. Figure 4-6 shows the results.

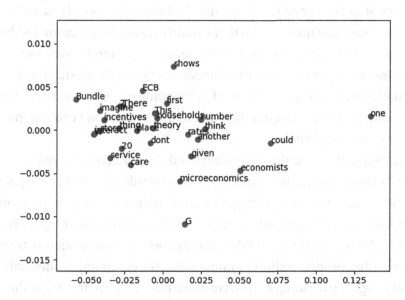

Figure 4-6. *CBoW word embedding visualization*

Global Vectors for Word Representation (GloVe)

GloVe is a contemporary and advanced method of vector representation of words. In 2014, Jeffrey Pennington, Richard Socher, and Christopher Manning wrote a paper in which they describe GloVe. This type of word embedding is an improvement over both matrix factorization–based representations of words, as well as the Skip-gram model. Matrix factorization–based methods of word representation are not particularly good at representing words with respect to their analogous nature. However, Skip-Gram and CBoW train on isolated windows of text and do not utilize the same information that a matrix-based factorization method does. Specifically, when we use LDA to create a topic model, we have to preprocess the text in a way that encodes each word with statistical information that represents the word in the context of the whole text. With Skip-Gram and CBoW, the one-hot encoded vector doesn't capture that same type of complexity.

GloVe specifically trains on "global word-to-word co-occurrence counts." Co-occurrence is the instance of two words appearing in a specific order alongside one another. By *global*, I mean the co-occurrence counts with respect to all documents in the corpus that we are analyzing. In this sense, GloVe is utilizing a bit of the intuition behind both models to try and overcome the respective shortcomings of the aforementioned alternatives.

Let's begin by defining a co-occurrence matrix, X. Each entry in the matrix represents the co-occurrence count of two specific words. More specifically, $X_{i,j}$ represents the number of times word j appears in the context of word i. The following notation is also worth noting:

$$\mathbf{X}_i = \sum_k X_{i,k} \qquad (4.4)$$

$$P_{i,j} = P(j|i) = \frac{X_{i,j}}{X_i} \qquad (4.5)$$

Equation 4.4 is defined as the number of times any word appears in the context of the word I. Equation 4.5 is the probability of a word j given word i. We define this probability as the co-occurrence account of word j appears in the context of the word "I" with the total co-occurrence counts of word i.

I suggest that the model should evaluate the ratios of co-occurrence probabilities, which we define as follows:

$$F\left(w_i, w_j, \tilde{w}_k\right) = \frac{P_{ik}}{P_{jk}} \tag{4.6}$$

$w \in \mathbb{R}^d$ = word vectors and $\tilde{w}_k \in \mathbb{R}^d$ = context vectors, $F = \exp(x)^*$

You should observe that our definition of F has an asterisk above it, particularly to indicate the fact that the value of F can be a multitude of things; however, we often derive it to be the preceding definition. The purpose of F is to encode the value yielded from the co-occurrence probabilities into the word embedding.

The following functions derive the target label and the error function we use to train the GloVe word embedding:

$$F\left(w_i^T \tilde{w}_k\right) = P_{i,k} \tag{4.7}$$

$$w_i^T \tilde{w}_j + b_i + \tilde{b}_j - \log X_{i,j} \tag{4.8}$$

$$J = \sum_{i,j=1}^{V} f\left(X_{i,j}\right)\left(w_i^T \tilde{w}_j + b_i + \tilde{b}_j - \log X_{i,j}\right)^2 \tag{4.9}$$

Where $f(X_{ij})$ = weighting function

As detailed in the GloVe paper, the weighting function should obey a few rules. Foremost, if f is a continuous function, it should vanish as $x \to 0$, $f(x)$ should be non-decreasing, and $f(x)$ should be relatively small for large values of x. These rules are to ensure that rare or frequent co-occurrence values are not overweighted in the training of the word embedding.

Although the weighting function can be altered, the GloVe paper suggests the following equation:

$$f(x) = \begin{cases} \left(\dfrac{x}{x_m}\right)^{\alpha} & \text{if } x < x_m \\ 1 & \text{otherwise} \end{cases}$$

x_m = maximum value of x, fixed to 100. The weighting function yields the values shown in Figure 4-7 with respect to the x value.

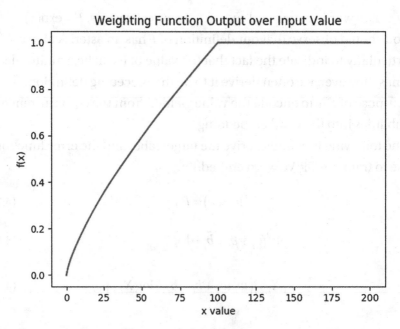

Figure 4-7. *Weighting function for GloVe*

Now that we have reviewed the model, it is useful for you to understand how to use pretrained word embeddings, particularly since not everyone will have the time or the ability to train these embeddings from scratch due to the difficult nature of acquiring all of this data. Although there is not necessarily one predetermined place to get a word embedding from, you should be aware of the following GitHub repository that contains

the files for multitudes of word embeddings: https://github.com/3Top/
word2vec-api#where-to-get-a-pretrained-models. You can feel free to
experiment and deploy these word embeddings for different tasks where
they see fit.

For this example, we will use a GloVe word embedding that contains
6 billion words and 50 features. This word embedding was trained from
data taken from Wikipedia and has a vocabulary containing 400,000 words.
Now, let's begin with the code, shown here:

```
def load_embedding(embedding_path='/path/to/glove.6B.50D.txt'):
    vocabulary, embedding = [], []
    for line in open(embedding_path, 'rb').readlines():
        row = line.strip().split(' ')
        vocabulary.append(row[0]), embedding.append(row[1:])
    vocabulary_length, embedding_dim = len(vocabulary),
    len(embedding[0])
    return vocabulary, np.asmatrix(embedding), vocabulary_
length, embedding_dim
```

We begin this problem by loading the word embeddings using the
native open() function to read the file line by line. Each line in the file
starts with a word in the vocabulary, and the subsequent entries in that
line represent the values within each of that word's vector. We iterate
through all the lines in the file, appending the word and the word vector to
their respective arrays. As such, we are able to create a list of words within
a vocabulary and reconstruct the word embeddings from a .txt file. This
trained embedding should look like Figure 4-8.

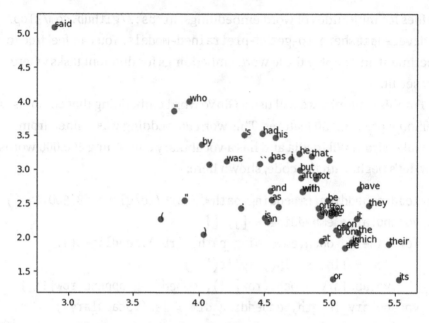

Figure 4-8. *GloVe pretrained embedding*

Figure 4-8 shows the representation of the first 50 words in the vocabulary when we look at the two principal components yielded from the transformation of our word embedding. Examples of words that seem to appear in similar contexts are *had* and *has, and,* and *as,* in addition to *his* and *he.* When comparing the cosine similarities of other words in the vocabulary, we observe the following.

```
Cosine Similarity Between so and u.s.: 0.5606769548631282
Cosine Similarity Between them and so: 0.8815159254335486
Cosine Similarity Between what and them: 0.8077565084355354
Cosine Similarity Between him and what: 0.7972281857691554
Cosine Similarity Between united and him: 0.5374600664967559
Cosine Similarity Between during and united: 0.6205250403136882
Cosine Similarity Between before and during: 0.8565694276984954
Cosine Similarity Between may and before: 0.7855322363492923
Cosine Similarity Between since and may: 0.7821437532357596
```

Example Problem 4.4: Using Trained Word Embeddings with LSTMs

Now that we have visually inspected the word embedding, let's focus on how to use trained embeddings with a deep learning algorithm. Let's imagine that we would like to include the following paragraph as additional training data for our word embedding.

```
sample_text = "'Living in different places has been the
greatest experience that I have had in my life. It has allowed
me to understand people from different walks of life, as well
as to question some of my own biases I have had with respect
to people who did not grow up as I did. If possible, everyone
should take an opportunity to travel somewhere separate from
where they grew up."'.replace('\n', ")
```

With our sample data assigned to a variable, let's begin by performing some of the same preprocessing steps that we have familiarized ourselves with, exemplified by the following body of code:

```
def sample_text_dictionary(data=_sample_text):
    count, dictionary = collections.Counter(data).most_
    common(), {} #creates list of word/count pairs;
    for word, _ in count:
        dictionary[word] = len(dictionary) #len(dictionary)
        increases each iteration
        reverse_dictionary = dict(zip(dictionary.values(),
        dictionary.keys()))
    dictionary_list = sorted(dictionary.items(),
    key = lambda x : x[1])
    return dictionary, reverse_dictionary, dictionary_list
```

We start by using a `remove_stop_words()` function, a redefinition of a sample preprocessing text algorithm defined in Chapter 3 that removes stop words from relatively straightforward sample data. When you are using data that isn't as clean as this sample data, I recommend that you preprocess the data in a manner similar to what you did using the economics textbook or *War and Peace*.

Moving to the `sample_text_dictionary()` function, we create a term frequency dictionary, and then return these variables. This process is important for you to understand, because this is an example of how we deal with words that are not in the vocabulary of a trained word embedding:

```
for i in range(len(dictionary)):
    word = dictionary_list[i][0]
    if word in vocabulary:
        _embedding_array.append(embedding_dictionary[word])
    else:
        _embedding_array.append(np.random.uniform(low=-0.2,
        high=0.2, size=embedding_dim))
```

We begin by creating a variable title: _embedding_array. This variable actually contains the word embedding representations of our sample text. To handle words that are not in the vocabulary, we will create a randomized distribution of numbers to simulate a word embedding, which we then feed as inputs to the neural network.

Moving forward, we make the final transformations to the embedding data before we create our computation graph.

```
embedding_array = np.asarray(_embedding_array)
decision_tree = spatial.KDTree(embedding_array, leafsize=100)
```

We will use a k-nearest neighbors tree to find the embedding that is closest to the array that our neural network outputs. From this, we use reverse_dictionary to find the word that matches the predicted embedding.

Let's build our computational graph, as follows:

```
#Initializing placeholders and other variables
X = tf.placeholder(tf.int32, shape=(None, None, n_input))
Y = tf.placeholder(tf.float32, shape=(None, embedding_dim))
weights = {'output': tf.Variable(tf.random_normal([n_hidden,
embedding_dim]))}
biases = {'output': tf.Variable(tf.random_normal([embedding_
dim]))}

_weights = tf.Variable(tf.constant(0.0, shape=[vocabulary_
length, embedding_dim]), trainable=True)
_embedding = tf.placeholder(tf.float32, [vocabulary_length,
embedding_dim])
embedding_initializer = _weights.assign(_embedding)
embedding_characters = tf.nn.embedding_lookup(_weights, X)
input_series = tf.reshape(embedding_characters, [-1, n_input])
input_series = tf.split(input_series, n_input, 1)
```

You will find most of this similar to the LSTM tutorial in Chapter 2, but direct your attention to the second grouping of code, specifically where we create the _weights and _embedding variables. When we are loading a trained word embedding, or have an embedding layer in our computational graph, the data must pass through this layer before it can get to the neural network. The dimension of the network is the number of words in the vocabulary by the number of features. Although the number of features when training one's own embedding can be altered, this is a predetermined value when we load a word embedding.

We assign the weights variable to the _embedding placeholder, which is ultimately the weights our optimizer is tuning, whereupon we create an embedding *characters variable.* The tf.nn.embedding_lookup() function specifically retrieves the index numbers of the _weights variable. Finally, we transform the embedding_characters variable into the input_series variable, which is actually directly fed into the LSTM layer.

From this point forward, the passage of data from the LSTM layer through the rest of the graph should be familiar from the tutorial. When executing the code, you should see output such as the following:

```
Input Sequence: ['me', 'to', 'understand', 'people']
Actual Label: from
Predicted Label: an
Epoch: 210
Error: 45.62042

Input Sequence: ['different', 'walks', 'of', 'life,']
Actual Label: as
Predicted Label: with
Epoch: 220
Error: 64.55679

Input Sequence: ['well', 'as', 'to', 'question']
Actual Label: some
Predicted Label: has
Epoch: 230
Error: 75.29771
```

An immediate suggestion for improving the error rate is to load different sample texts, perhaps from an actual corpus of data to train on, as the limited amount of data does not allow the accuracy to improve significantly much.

Another suggestion is to use the `load_data()` function that is commented out loading your own PDF file and experimenting from that point forward.

Now that we have reviewed the methods in which we can represent words as vectors, let's discuss other textual representations. Thankfully, since most of these are Word2Vec abstractions , it will not require nearly as much explanation this time around.

Paragraph2Vec: Distributed Memory of Paragraph Vectors (PV-DM)

Paragraph2Vec is an algorithm that allows us to represent objects of varying length, from sentences to whole documents, for the same purposes that we represented words as vectors in the previous examples. This technique was developed by Quoc Le and Tomas Mikolov, and largely is based off the Word2Vec algorithm.

In Paragraph2Vec, we represent each paragraph as a unique vector in a matrix, D. Every word is also mapped to a unique vector, represented by a column in matrix W. We subsequently construct a matrix, h, which is formed by concatenating matrices W and D. We think of this paragraph token as an analog to the cell state from the LSTM, in that it is providing memory to the current context in the form of the topic of the paragraph. Intuitively, that means that matrix W is the same across all paragraphs, such that we observe the same representation of a given word. Training occurs as it does in Word2Vec, and the negative sampling can occur in this instance by sampling from a fixed-length context in a random paragraph.

To ensure that you understand how this works functionally, let's look at one final example in this chapter.

Example Problem 4.5: Paragraph2Vec Example with Movie Review Data

Once again, Gensim thankfully has a Doc2Vec method that makes implementation of this algorithm relatively straightforward. In this example, we will keep things relatively simple and represent sentences in a vector space, rather than create or approximate a paragraph tokenizer, which we would likely want to be more precise than a heuristic that would be relatively quick to draw up (i.e., paragraphs comprised of four sentences each). In the doc2vec_example.py file, there are only slight differences in the Doc2Vec model and the Word2Vec model, specifically the preprocessing.

```
def gensim_preprocess_data(max_pages):
    sentences = namedtuple('sentence', 'words tags')
    _sentences = sent_tokenize(load_data(max_pages=max_pages))
    documents = []
    for i, text in enumerate(_sentences):
        words, tags = text.lower().split(), [i]
        documents.append(sentences(words, tags))
    return documents
```

The Doc2Vec implementation expects what is known as a named *tuple object*. This tuple contains a list of tokenized words contained in the sentence, as well as an integer that indexes this document. In online documentation, some people utilize a class object entitled LabledLineSentence(); however, this performs the necessary preprocessing the same way. When we run our script, we iterate through all the sentences that we are analyzing, and view their associated cosine similarities. The following is an example of some of them:

```
Document sentence(words=['this', 'text', 'adapted', 'the',
'saylor', 'foundation', 'creative', 'commons', 'attribution-
```

```
noncommercial-sharealike', '3.0', 'license', 'without',
'attribution', 'requested', 'works', 'original', 'creator',
'licensee', '.'], tags=[0])
```

```
Document sentence(words=['saylor', 'url', ':', 'http', ':',
'//www.saylor.org/books', 'saylor.org', '1', 'preface', 'we',
'written', 'fundamentally', 'different', 'text', 'principles',
'economics', ',', 'based', 'two', 'premises', ':', '1', '.'],
tags=[1])
```

```
Cosine Similarity Between Documents: -0.025641936104727547
Document sentence(words=['saylor', 'url', ':', 'http', ':',
'//www.saylor.org/books', 'saylor.org', '1', 'preface', 'we',
'written', 'fundamentally', 'different', 'text', 'principles',
'economics', ',', 'based', 'two', 'premises', ':', '1', '.'],
tags=[1])
```

```
Document sentence(words=['students', 'motivated', 'study',
'economics', 'see', 'relates', 'lives', '.'], tags=[2])
```

```
Cosine Similarity Between Documents:
0.06511943195883922
```

Beyond this, Gensim also allows us to infer vectors without having to retrain our models on these vectors. This is particularly important in Chapter 5, where we apply word embeddings in a practical setting. You can see this functionality when we execute the code with the training_ example parameter set to False. We have two sample documents, which we define at the beginning of the file:

```
sample_text1 = "'I love italian food. My favorite items are
pizza and pasta, especially garlic bread. The best italian food
I have had has been in New York. Little Italy was very fun"'
```

```
sample_text2 = "'My favorite time of italian food is pasta with
alfredo sauce. It is very creamy but the cheese is the best
part. Whenevr I go to an italian restaurant, I am always
certain to get a plate."'
```

These two examples are fairly similar. When we train our model—more than 300 pages worth of data from an economics textbook, we get the following results:

```
 Cosine Similarity Between Sample Texts:
0.9911814256706748
```

Again, you should be aware that they will likely need significantly larger amounts of data to get reasonable results across unseen data. These examples show them how to train and infer vectors using various frameworks. For those who are dedicated to training their own word embeddings, the path forward should be fairly clear.

Summary

Before we move on to work on natural language processing tasks, let's recap some of the most important things learned in this chapter. As you saw in Chapter 3, preprocessing data correctly is the majority of the work that we need to perform when applying deep learning to natural language processing. Beyond cleaning out stop words, punctuation, and statistical noise, you should be prepared to wrangle data and organize it in an interpretable format for the neural network. Well-trained word embeddings often require the collection of billions of tokens.

Making sure that you aggregate the right data is extremely important, as a couple billion tokens from radically different data sources can leave you with an embedding that doesn't yield much of anything useful. Although some of our examples yielded positive results, it does not mean these applications would work in a production environment. You must (responsibly) collect large amounts of text data from sources while maintaining homogeneity in vocabulary and context.

In the following chapter, we conclude the book by working on applications of recurrent neural networks.

CHAPTER 5

Text Generation, Machine Translation, and Other Recurrent Language Modeling Tasks

In Chapter 4, I introduced you to some of the more advanced deep learning and NLP techniques, and I discussed how to implement these models in some basic problems, such as mapping word vectors. Before we conclude this book, I will discuss a handful of other NLP tasks that are more domain-specific, but nonetheless useful to go through.

By this point, you should be relatively comfortable with preprocessing text data in various formats, and you should understand a few NLP tasks, such as document classification, well enough to perform them. As such, this chapter focuses on combining many of the skills we have worked with by tackling a couple of problems. All solutions provided in this chapter are feasible. You are more than welcome to present or complete new solutions that outperform them.

© Taweh Beysolow II 2018
T. Beysolow II, *Applied Natural Language Processing with Python*,
https://doi.org/10.1007/978-1-4842-3733-5_5

Text Generation with LSTMs

Text generation is increasingly an important feature in AI-based tools. Particularly when working with large amounts of data, it is useful for systems to be able to communicate with users to provide a more immersive and informative experience. For text generation, the main goal is to create a generative model that provides some sort of insight with respect to the data. You should be aware that text generation should not necessarily create a summary of the document, but generate an output that is descriptive of the input text. Let's start by inspecting the problem.

Initially, for such a task, we need a data source. From that, our data source changes the results. For this task, we start by working with *Harry Potter and the Sorcerer's Stone*. I chose this book since the context should provide some fairly notable results with respect to the topics that are contained within the generated text.

Let's go through the steps that we've become accustomed to. We will utilize the `load_data()` preprocessing function that we used in `word_embeddings.py`; however, the only change that we will make is loading `harry_potter.pdf` instead of `economics_textbook.pdf`.

That said, this function allows you to easily utilize the preprocessing function for whatever purpose, so long as the directory and other arguments are changed. Being that this is a text generation example, we should not clean the data beyond removing non-ASCII characters.

The following is an example of how the data appears:

```
"Harry Potter Sorcerer's Stone CHAPTER ONE THE BOY WHO LIVED
Mr. Mrs. Dursley, number four, Privet Drive, proud say
perfectly normal, thank much. They last people 'd expect
involved anything strange mysterious, n't hold nonsense. Mr.
Dursley director firm called Grunnings, made drills. He big,
beefy man hardly neck, although large mustache. Mrs. Dursley
thin blonde nearly twice usual amount neck, came useful spent
```

much time craning garden fences, spying neighbors. The Dursleys
small son called Dudley opinion finer boy anywhere. The
Dursleys everything wanted, also secret, greatest fear somebody
would discover. They think could bear anyone found Potters.
Mrs. Potter Mrs. Dursley's sister, n't met several years; fact,
Mrs. Dursley pretended n't sister, sister good-for-nothing
husband unDursleyish possible. The Dursleys shuddered think
neighbors would say Potters arrived street. The Dursleys knew
Potters small son,, never even seen. This boy another good
reason keeping Potters away; n't want Dudley mixing child like.
When Mr. Mrs. Dursley woke dull, gray Tuesday story starts,
nothing cloudy sky outside suggest strange mysterious things
would soon happening country. Mr. Dursley hummed picked boring
tie work, Mrs. Dursley gossiped away happily wrestled screaming
Dudley high chair. None noticed large, tawny owl flutter past
window. At half past eight, Mr. Dursley picked briefcase,
pecked Mrs. Dursley cheek, tried kiss Dudley good-bye missed, 1
Dudley tantrum throwing cereal walls. `` Little tyke, "chortled
Mr. Dursley left house. He got car backed number four's drive.
It corner street noticed first sign something peculiar -- cat
reading map. For second, Mr. Dursley n't realize seen -- jerked
head around look. There tabby cat standing corner Privet Drive,
n't map sight. What could thinking ? It must trick light. Mr.
Dursley blinked stared cat. It stared back. As Mr. Dursley
drove around corner road, watched cat mirror. It reading sign
said Privet Drive --, looking sign; cats..."

Let's inspect our preprocessing function.

```
def preprocess_data(sequence_length=sequence_length, max_
pages=max_pages, pdf_file=pdf_file):
    text_data = load_data(max_pages=max_pages, pdf_file=pdf_file)
    characters = list(set(text_data.lower()))
```

```
character_dict = dict((character, i) for i, character in
enumerate(characters))
int_dictionary = dict((i, character) for i, character in
enumerate(characters))
num_chars, vocab_size = len(text_data), len(characters)
x, y = [], []

for i in range(0, num_chars - sequence_length, 1):
    input_sequence = text_data[i: i+sequence_length]
    output_sequence = text_data[i+sequence_length]
    x.append([character_dict[character.lower()] for
    character in input_sequence])
    y.append(character_dict[output_sequence.lower()])

for k in range(0, len(x)): x[i] = [_x for _x in x[i]]
x = np.reshape(x, (len(x), sequence_length, 1))
x, y = x/float(vocab_size), np_utils.to_categorical(y)
return x, y, num_chars, vocab_size, character_dict,
int_dictionary
```

When inspecting the function, we use methods similar to the tf_
preprocess_data() function in the toy example of a Skip-Gram model.
Our input and output sequences are fixed lengths, and we will transform
the y variable to a one-hot encoded vector, where each entry in the vector
represents a possible character. We represent the sequence of characters
as a matrix, where each row represents the entire observation and each
column represents a character.

Let's look at the first example of Keras code used in the book.

```
def create_rnn(num_units=num_units, activation=activation):
    model = Sequential()
    model.add(LSTM(num_units, activation=activation,
    input_shape=(None, x.shape[1])))
    model.add(Dense(y.shape[1], activation='softmax'))
```

```
model.compile(loss='categorical_crossentropy',
optimizer='adam')
model.summary()
return model
```

Keras, unlike TensorFlow, is considerably less verbose. As such, this makes changing the architecture of a model relatively easy. We instantiate a model by assigning it to a variable, and then simply add layer types with the Sequential().add() function.

After running the network with 200 epochs, we get the following result:

```
driv, proud say perfecdly normal, thanp much. they last
people 'd expect involved anytsing strange mysterious, s't
hold donsense. mr. dursley director firm called grunnings,
made drills. he big, berfy man, ardly neck, althougl larte
mustache. mrs. dursley thic -londe. early twece uiual amount
necd, came ueeful spent much time craning geddon fences,
spying neighbors. the dursleys small son called dudley
opinion finer boy anyw  rd. the dursleys everything wanted,
slso secret, greatest fear somebody would discover. they
thinn could bear antone found potters. mrs. potterimrs.
dursley's sister, n't met several years; fact, mrs. dursley
pretended n't sister, sister good-sur-notding husband
undursleyir  pousible. the dursleys suuddered think auigybors
would say potters. arrived strett. the dursleys knew potters
small. on,  ever even seen. thit boy another good reason
keeping potters away; n'e want dudley mixing child like.
wten mr. mrs. dursley woke dull, gray tuesday story startss,
nothing cloudy skycoutside suggest strange mytter ous taings
would soon darpening codntry. mr. dursley tummed picked
boring tie work, mrs. dursley gosudaed away happily wrestled
screaming dudley aigh cuair. noneoloticed large, tawny owl
flutter past wincow. at ialf past, ight, mr. dursley picked
```

briefcase, pecked mrs. dursley cheek, tried kiss dudley good-
bye missed, 1 dudley tantrum,hrowigg cereal walls. `` lwttle
tykp, "chortled mr. dursley left house. he got car backel
number four's drive. it corner street noticed fir t sign
somathing pcculilr -- cat feading,ap. for sicond, mr. dursley
r't realize scen -- jerked head around look. thereytab y
cat standing corneraprivet drive, n'tamap sight. what sould
thinking ? it muse trick light. mr. dursley blinked stared
cat. it stayed back. as mr. dursley drove around corner road,
watched catcmirror. it reading sign saidsprivet druve --,
lookingtsign; cats could n't read maps signs. mr. durs

Note   Some of the text is interpretable, but obviously not everything
is as good as it could be. In this instance, I suggest that you allow the
neural network to train longer and to add more data. Also consider
using different models and model architectures. Beyond this example,
it would be useful to present a more advanced version of the LSTM
that is also useful for speech modeling.

Bidirectional RNNs (BRNN)

BRNNs were created in 1997 by Mike Schuster and Kukdip Paliwal, who
introduced the technique to a signal-processing academic journal. The
purpose of the model was to utilize information moving in both a "positive
and negative time direction." Specifically, they wanted to utilize both
the information moving up to the prediction, as well as the same stream
of inputs moving in the opposite direction. Figure 5-1 illustrates the
architecture of a BRNN.

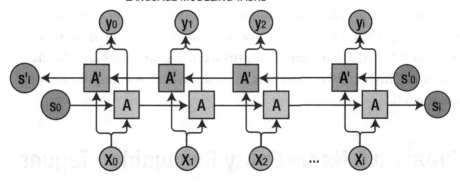

Figure 5-1. *Bidirectional RNN*

Let's imagine that that we have a sequence of words, such as the following: The man walks down the boardwalk.

In a regular RNN, assuming that we wanted to predict the word *boardwalk*, the input data would be *The, man, walks, down, the.* If we input bigrams, it would be *The, man, man, walks,* and so forth. We keep moving through the input data, predicting the word that is most likely to come next at each time step, culminating in our final target label, a probability that corresponds to the one-hot encoded vector that is most likely to be present given the input data. The only difference in a BRNN is that while we are predicting the sequence left-to-right, we also are predicting the sequence right-to-left.

BRNNs have been particularly useful for NLP tasks. The following is the code for building a BRNN:

```
def create_lstm(input_shape=(1, x.shape[1])):
        model = Sequential()
        model.add(Bidirectional(LSTM(unites=n_units,
                                    activation=activation),
                                    input_shape=input_shape))

        model.add(Dense(train_y.shape[1]), activation=out_act)
        model.compile(loss='categorical_crossentropy',
        metrics=['accuracy'])
        return model
```

The structure of the bidirectional RNN is nearly identical in that we are only adding a `Bidirectional()` cast over our layer. This often increases the time it takes to train neural networks, but in general, it outperforms traditional RNN architectures in many tasks. With this in mind, let's apply our model.

Creating a Name Entity Recognition Tagger

People who have worked with NLTK or similar packages have likely come across the *name entity recognition* (NER) tagger. NER taggers typically output a label that identifies the entity within larger categories (person, organization, location, etc.). Creating an NER tagger requires a large amount of annotated data.

For this task, we will use a data set from Kaggle. When we unzip the data, we see that it comes in the following format:

played	on	Monday	(	home	team	in	CAPS	)	:
VBD	IN	NNP	(	NN	NN	IN	NNP	)	:
O	O	O	O	O	O	O	O	O	O
American	League								
NNP	NNP								
B-MISC	I-MISC								
Cleveland	2	DETROIT	1						
NNP	CD	NNP	CD						
B-ORG	O	B-ORG	O						
BALTIMORE	12	Oakland	11	(	10	innings			)
VB	CD	NNP	CD	(	CD	NN			)
B-ORG	O	B-ORG	O	O	O	O	O		
TORONTO	5	Minnesota	3						
TO	CD	NNP	CD						
B-ORG	O	B-ORG	O						
Milwaukee	3	CHICAGO	2						

NNP	CD	NNP	CD
B-ORG	O	B-ORG	O
Boston	4	CALIFORNIA	1

The data is tab-delimited but also in .txt format. This requires some data wrangling before we get to training the BRNN.

Let's start by turning the text data into an interpretable format, as follows:

```
def load_data():
    text_data = open('/Users/tawehbeysolow/Downloads/train.
    txt', 'rb').readlines()
    text_data = [text_data[k].replace('\t', ' ').split() for k
    in range(0, len(text_data))]
    index = range(0, len(text_data), 3)

    #Transforming data to matrix format for neural network
    input_data = list()
    for i in range(1, len(index)-1):
        rows = text_data[index[i-1]:index[i]]
        sentence_no = np.array([i for i in np.repeat(i,
        len(rows[0]))], dtype=str)
        rows.append(np.array(sentence_no))
        rows = np.array(rows).T
        input_data.append(rows)
```

We must first iterate through each line of the .txt file. Notice that the data is organized in groups of three. A typical grouping looks like the following:

```
text_data[0]
['played', 'on', 'Monday', '(', 'home', 'team', 'in', 'CAPS',
')', ':']
 text_data[1]
```

```
['VBD', 'IN', 'NNP', '(', 'NN', 'NN', 'IN', 'NNP', ')', ':']
text_data[2]
['0', '0', '0', '0', '0', '0', '0', '0', '0', '0']
```

The first set of observations contains the text itself, the second set of observations contains the name entity tag, and the final set contains the specific tag. Back to the preprocessing function, we take the groupings of sentences and append an array that contains a sentence number label, which I will discuss the importance of shortly.

When looking at a snapshot of the input_data variable, we see the following:

```
input_data[0:1]
[array([['played', 'VBD', '0', '1'],
       ['on', 'IN', '0', '1'],
       ['Monday', 'NNP', '0', '1'],
       ['(', '(', '0', '1'],
       ['home', 'NN', '0', '1'],
       ['team', 'NN', '0', '1'],
       ['in', 'IN', '0', '1'],
       ['CAPS', 'NNP', '0', '1'],
       [')', ')', '0', '1'],
       [':', ':', '0', '1']], dtype='|S6')]
```

We need to remove the sentence label while observing the data in such a fashion that the neural network implicitly understands how these sentences are grouped. The reason we want to remove this label is that neural networks read categorical labels (which the sentence number is an analog for) in such a way that higher-numbered sentences explicitly have a greater importance than lower-numbered sentences. For this task, I assume most you understand we do *not* want to bake this into the training process. As such, we move to the following body of code:

```
input_data = pan.DataFrame(np.concatenate([input_data[j] for
j in range(0,len(input_data))]),
                    columns=['word', 'pos', 'tag', 'sent_no'])
```

```
labels, vocabulary = list(set(input_data['tag'].values)),
list(set(input_data['word'].values))
vocabulary.append('endpad'); vocab_size = len(vocabulary);
label_size = len(labels)
```

```
aggregate_function = lambda input: [(word, pos, label) for
word, pos, label in zip(input['word'].values.tolist(),
  input['pos'].values.tolist(),
  input['tag'].values.tolist())]
```

We organize the input_data into a data frame, and then create a
couple of other variables that we will use in the function later, as well as
the train_brnn_keras() function. Some of these variables are familiar to
others present in the scripts from the prior chapter (vocab_size represents
the number of words in the vocabulary, for example). However, the
important parts are mainly the last two variables, which is what you should
focus on to solve this problem.

The lambda function, aggregate_function, takes a data frame as
an input, and then returns a three-tuple for each observation within a
grouping. This is precisely how we will group all the observations within
one sentence. A snapshot of our data after this transformation yields the
following:

```
sentences[0]
[('played', 'VBD', 'O'), ('on', 'IN', 'O'), ('Monday', 'NNP',
'O'), ('(', '(', 'O'), ('home', 'NN', 'O'), ('team', 'NN',
'O'), ('in', 'IN', 'O'), ('CAPS', 'NNP', 'O'), (')', ')', 'O'),
(':', ':', 'O')]
```

We have nearly finished all the necessary preprocessing; however, there is a key step that you should be aware of.

```
x = [[word_dictionary[word[0]] for word in sent] for sent
in sentences]
x = pad_sequences(maxlen=input_shape, sequences=x,
padding='post', value=0)
y = [[label_dictionary[word[2]] for word in sent] for sent
in sentences]
y = pad_sequences(maxlen=input_shape, sequences=y,
padding='post', value=0)
 = [np_utils.to_categorical(label, num_classes=label_size)
for label in y]
```

In the preceding lines of code, we are transforming our words to their integer labels as we did in many other examples, and creating a one-hot encoded matrix. This is similar to the previous chapter, however, we should specifically not use the pad_sequences() function.

When working with sentence data, we do not always get sentences of equal length; however, the input matrix for the neural network has to have an equal number of features across all observations. *Zero padding* is used to add the extra features that normalize the size of all observations.

With this step done, we are now ready to move to training our neural network. Our model is as follows:

```
def create_brnn():
        model = Sequential()
        model.add(Embedding(input_dim=vocab_size+1,
        output_dim=output_dim,
                               input_length=input_shape,
                               mask_zero=True))
        model.add(Bidirectional(LSTM(units=n_units,
        activation=activation,
                                     return_sequences=True)))
```

```
model.add(TimeDistributed(Dense(label_size,
activation=out_act)))
model.compile(optimizer='adam', loss='categorical_
crossentropy', metrics=['accuracy'])
model.summary()
return model
```

Most of our model is similar to the prior Keras models built in this chapter; however, we have an embedding layer (analogous to a word embedding) that is stacked on top of the bidirectional LSTM, which is subsequently staked on top of a fully connected output layer.

We train our network on roughly 90% of the data we have, and then subsequently evaluate the results. We find that our tagger on the training data yields an accuracy of 90% and higher, depending on the number of epochs we train it for.

Now that we have dealt with this classification task and sufficiently worked with BRNNs, let's move on to another neural network model and discuss how it can be effectively applied to another NLP task.

Sequence-to-Sequence Models (Seq2Seq)

Sequence-to-sequence models (seq2seq) are notable because they take in an input sequence and return an output sequence, both of variable length. This makes this model particularly powerful, and it is predisposed to perform well on language modeling tasks. The particular model that we will utilize is best summarized in a paper by Sutskever et al. Figure 5-2 illustrates the model.

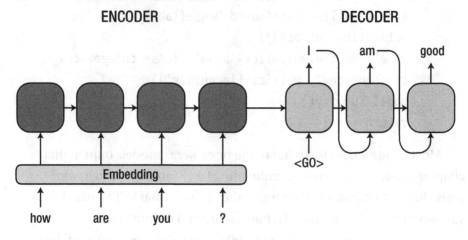

Figure 5-2. *Encoder-decoder model*

The model is generally comprised of two parts: an encoder and a
decoder. Both the encoder and the decoder are RNNs. The encoder reads
the input sequence and outputs a fixed-length vector in addition to the
hidden and cell states from the LSTM unit. Subsequently, the decoder
takes this fixed-length vector, in addition to the output hidden and cell
states, and uses them as inputs to the first of its LSTM units. The decoder
outputs a fixed-length vector, which we will evaluate as a target label. We
will perform prediction one character at a time, which easily allows us to
evaluate sequences of varying length from one observation to the next.
Next, you see this model in action.

Question and Answer with Neural Network Models

One popular application of deep learning to NLP is the chatbot. Many
companies use chatbots to handle generic customer service requests,
which require them to be flexible in translating questions into answers.
While the test case that we look at is a microcosm of questions and

answers, it is an example of how we can train a neural network to properly answer a question. We will use the Stanford Question Answering Dataset. Although it is more representative of general knowledge, you would do well to recognize the way in which these problems are structured.

Let's begin by examining how we will preprocess the data by utilizing the following function:

```
dataset = json.load(open('/Users/tawehbeysolow/Downloads/
qadataset.json', 'rb'))['data']
questions, answers = [], []
for j in range(0, len(dataset)):
    for k in range(0, len(dataset[j])):
        for i in range(0, len(dataset[j]['paragraphs'][k]
        ['qas'])):
            questions.append(remove_non_ascii(dataset[j]
            ['paragraphs'][k]['qas'][i]['question']))

            answers.append(remove_non_ascii(dataset[j]
            ['paragraphs'][k]['qas'][i]['answers'][0]
            ['text']))
```

When we look at a snapshot of the data, we observe the following structure:

```
[{u'paragraphs': [{u'qas': [{u'question': u'To whom did the
Virgin Mary allegedly appear in 1858 in Lourdes France?',
u'id': u'5733be284776f41900661182', u'answers': [{u'text':
u'Saint Bernadette Soubirous', u'answer_start': 515}]},
{u'question': u'What is in front of the Notre Dame Main
Building?', u'id': u'5733be284776f4190066117f', u'answers':
[{u'text': u'a copper statue of Christ', u'answer_start':
188}]}, {u'question': u'The Basilica of the Sacred heart
at Notre Dame is beside to which structure?', u'id':
u'5733be284776f41900661180', u'answers': [{u'text': u'the Main
Building', u'answer_start': 279}]}, {u'question': u'What is the
```

Grotto at Notre Dame?', u'id': u'5733be284776f41900661181',
u'answers': [{u'text': u'a Marian place of prayer and
reflection', u'answer_start': 381}]}, {u'question': u'What
sits on top of the Main Building at Notre Dame?', u'id':
u'5733be284776f4190066117e', u'answers': [{u'text': u'a
golden statue of the Virgin Mary', u'answer_start': 92}]}],
u'context': u'Architecturally, the school has a Catholic
character. Atop the Main Building\'s gold dome is a golden
statue of the Virgin Mary. Immediately in front of the Main
Building and facing it, is a copper statue of Christ with arms
upraised with the legend "Venite Ad Me Omnes". Next to the
Main Building is the Basilica of the Sacred Heart. Immediately
behind the basilica is the Grotto, a Marian place of prayer and
reflection. It is a replica of the grotto at Lourdes, France
where the Virgin Mary reputedly appeared to Saint Bernadette
Soubirous in 1858. At the end of the main drive (and in a
direct line that connects through 3 statues and the Gold
Dome), is a simple, modern stone statue of Mary.'}, {u'qas':
[{u'question': u'When did the Scholastic Magazine of Notre
dame begin publishing?', u'id': u'5733bf84d058e614000b61be',
u'answers'

We have a JSON file with question and answers. Similar to the name entity recognition task, we need to preprocess our data into a matrix format that we can input into a neural network. We must first collect the questions that correspond to the proper answers. Then we iterate through the JSON file, and append each of the questions and answers to the corresponding arrays.

Now let's discuss how we are actually going to frame the problem for the neural network. Rather than have the neural network predict each word, we are going to have the neural network predict each character given an input sequence of characters. Since this is a multilabel classification problem, we will output a softmax probability for each element of the output vector, and then choose the vector with the highest probability. This represents the character that is most likely to proceed given the prior input sequence.

After we have done this for the entire output sequence, we will concatenate this array of outputted characters so that we get a human-readable message. As such, we move forward to the following part of the code:

```
input_chars, output_chars = set(), set()

for i in range(0, len(questions)):
    for char in questions[i]:
        if char not in input_chars: input_chars.add(char.
        lower())

for i in range(0, len(answers)):
    for char in answers[i]:
        if char not in output_chars: output_chars.add(char.
        lower())

input_chars, output_chars = sorted(list(input_chars)),
sorted(list(output_chars))
n_encoder_tokens, n_decoder_tokens = len(input_chars),
len(output_chars)
```

We iterated through each of the questions and answers, and collected all the unique individual characters in both the output and input sequences. This yields the following sets, which represent the input and output characters, respectively.

```
input_chars; output_chars
[u' ', u'"', u'#', u'%', u'&', u'"!", u'(', u')', u',', u'-',
u'.', u'/', u'0', u'1', u'2', u'3', u'4', u'5', u'6', u'7',
u'8', u'9', u':', u';', u'>', u'?', u'_', u'a', u'b', u'c',
u'd', u'e', u'f', u'g', u'h', u'i', u'j', u'k', u'l', u'm',
u'n', u'o', u'p', u'q', u'r', u's', u't', u'u', u'v', u'w',
u'x', u'y', u'z']
```

```
[u' ', u'!', u'"', u'$', u'%', u'&', u"'", u'(', u')', u'+',
u',', u'-', u'.', u'/', u'0', u'1', u'2', u'3', u'4', u'5',
u'6', u'7', u'8', u'9', u':', u';', u'?', u'[', u']', u'a',
u'b', u'c', u'd', u'e', u'f', u'g', u'h', u'i', u'j', u'k',
u'l', u'm', u'n', u'o', u'p', u'q', u'r', u's', u't', u'u',
u'v', u'w', u'x', u'y', u'z']
```

The two lists contain 53 and 55 characters, respectively; however, they are virtually homogenous and contain all the letters of the alphabet, plus some grammatical and numerical characters.

We move to the most important part of the preprocessing, in which we transform our input sequences to one-hot encoded vectors that are interpretable by the neural network.

```
(code redacted, please see github)
    x_encoder = np.zeros((len(questions), max_encoder_len,
    n_encoder_tokens))
    x_decoder = np.zeros((len(questions), max_decoder_len,
    n_decoder_tokens))
    y_decoder = np.zeros((len(questions), max_decoder_len,
    n_decoder_tokens))

    for i, (input, output) in enumerate(zip(questions,
    answers)):
        for _character, character in enumerate(input):
            x_encoder[i, _character, input_
            dictionary[character.lower()]] = 1.

        for _character, character in enumerate(output):
            x_decoder[i, _character, output_
            dictionary[character.lower()]] = 1.

            if i > 0: y_decoder[i, _character,
            output_dictionary[character.lower()]] = 1.
```

We start by instantiating two input vectors and an output vector,
denoted by x_encoder, x_decoder, and y_encoder. Sequentially, this
represents the order in which the data passes through the neural network
and validated against the target label. While the one-hot encoding that
we chose to create here is similar, we make a minor change by creating a
three-dimensional array to evaluate each question and answer. Each row
represents a question, each time step represents a character, and each
column represents the type of character within our set of characters. We
repeat this process for each question-and-answer pair until we have an
array with the entire data set, which yields 4980 observations of data.

The last step defines the model, as given by the encoder_decoder()
function.

```
def encoder_decoder(n_encoder_tokens, n_decoder_tokens):

    encoder_input = Input(shape=(None, n_encoder_tokens))
    encoder = LSTM(n_units, return_state=True)
    encoder_output, hidden_state, cell_state = encoder(encoder_
    input)
    encoder_states = [hidden_state, cell_state]

    decoder_input = Input(shape=(None, n_decoder_tokens))
    decoder = LSTM(n_units, return_state=True,
    return_sequences=True)
    decoder_output, _, _ = decoder(decoder_input,
    initial_state=encoder_states)

    decoder = Dense(n_decoder_tokens, activation='softmax')
    (decoder_output)
    model = Model([encoder_input, decoder_input], decoder)
    model.compile(optimizer='adam', loss='categorical_
    crossentropy', metrics=['accuracy'])
    model.summary()
    return model
```

We instantiated our model slightly differently than other Keras models. This method of creating a model is done through using the Functional API, rather than relying on the sequential model, as we have often done. Specifically, this method is useful when creating more complex models, such as seq2seq models, and is relatively straightforward once you have learned how to use the sequential model. Rather than adding layers to the sequential model, we instantiate different layers as variables and then pass the data by calling the tensor we created. We see this when observing the encoder_output variable when we instantiate it by calling encoder(encoder_input). We keep doing this through the encoder-decoder phase until we reach an output vector, which we define as a *dense/fully connected layer* with a softmax activation function.

Finally, we move to training, and observe the following results:

Model Prediction: saint bernadette soubiroust

Actual Output: saint bernadette soubirous

Model Prediction: a copper statue of christ

Actual Output: a copper statue of christ

Model Prediction: the main building

Actual Output: the main building

Model Prediction: a marian place of prayer and reflection

Actual Output: a marian place of prayer and reflection

Model Prediction: a golden statue of the virgin mary

Actual Output: a golden statue of the virgin mary

Model Prediction: september 18760

Actual Output: september 1876

Model Prediction: twice

```
Actual Output: twice

Model Prediction: the observer

Actual Output: the observer

Model Prediction: three

Actual Output: three

Model Prediction: 19877
Actual Output: 1987
```

As you can see, this model performs considerably well, with only three epochs. Although there are some problems with the spelling from adding extra characters, the messages themselves are correct in most instances. Feel free to keep experimenting with this problem, particularly by altering the model architecture to see if there is one that yields better accuracy.

Summary

With the chapter coming to a close, we should review the concepts that are most important in helping us successfully train our algorithms. Primarily, you should take note of the model types that are appropriate for different problems. The encoder-decoder model architecture introduces the "many-to-many" input-output scheme and shows where it is appropriate to apply it.

Secondarily, you should take note of where preprocessing techniques can be applied to seemingly different but related problems. The translation of data from one language to another uses the same preprocessing steps as creating a neural network that answered questions based on different responses. Paying attention to these modeling steps and how they relate to the underlying structure of the data can save you time on seemingly innocuous tasks.

Conclusion and Final Statements

We have reached the end of this book. We solved a wide array of NLP problems of varying complexities and domains. There are many concepts that are constant across all problem types, most specifically data preprocessing. The vast majority of what makes machine learning difficult is preprocessing data. You saw that similar problem types share preprocessing steps, as we often reused parts of solutions as we moved to more advanced problems.

There are some final principles that are worth remembering from this point forward. NLP with deep learning can require large amounts of text data. Collect it carefully and responsibly, and consider your options when dealing with large data sets with respect to choice of language for optimized run time (C/C++ vs. Python, etc.).

Neural networks, by and large, are fairly straightforward models to work with. The difficulty is finding good data that has predictive power, in addition to structuring it in such a way that our neural network can find patterns to exploit.

Study carefully the preprocessing steps to take for document classification, creating a word embedding, or creating an NER tagger, for example. Each of these represents feature extraction schemes that can be applied to different problems and illuminate a path forward during your research.

Although intelligent preprocessing of data spoken about fairly often in the machine learning community, it is particularly true of the NLP paradigm of deep learning and data science. The models that we have trained give you a roadmap on how to work with similar data sets in professional or academic environments. However, this does not mean that the models we have deployed could be used in production and work well.

There are a considerable number of variables that I did not discuss, being that they are problems of maintaining production systems rather than the theory behind a model. Examples include unknown words in vocabularies that appear over time, when to retrain models, how to evaluate multiple models' outputs simultaneously, and so forth.

In my experience, finding out when to retrain models has best been solved by collecting large amounts of live performance data. See when signals deprecate, if they do at all, and track the effect of retraining, as well as the persistence in retraining of models. Even if your model is accurate, it does not mean that it will be easy to use in practice.

Think carefully about how to handle false classifications, particularly if the penalty for misclassification could cause the loss of money and/or other resources. Do not be afraid to utilize multiple models for multiple problem types. When experimenting, start simple and gradually add complexity as needed. This is significantly easier than trying to design something very complex in the beginning and then trying to debug a system that you do not understand.

You are encouraged to reread this book at your leisure, as well as for reference, in addition to utilizing the code on my GitHub page to tackle the problems in their own unique fashion. While reading this book provides a start, the only way to become proficient in data science is to practice the problems on your own.

I hope you have enjoyed learning about natural language processing and deep learning as much as I have enjoyed explaining it.

Index

© Taweh Beysolow II 2018
T. Beysolow II, *Applied Natural Language Processing with Python*,
https://doi.org/10.1007/978-1-4842-3733-5

W, X, Y

Z

Printed in the United States
By Bookmasters